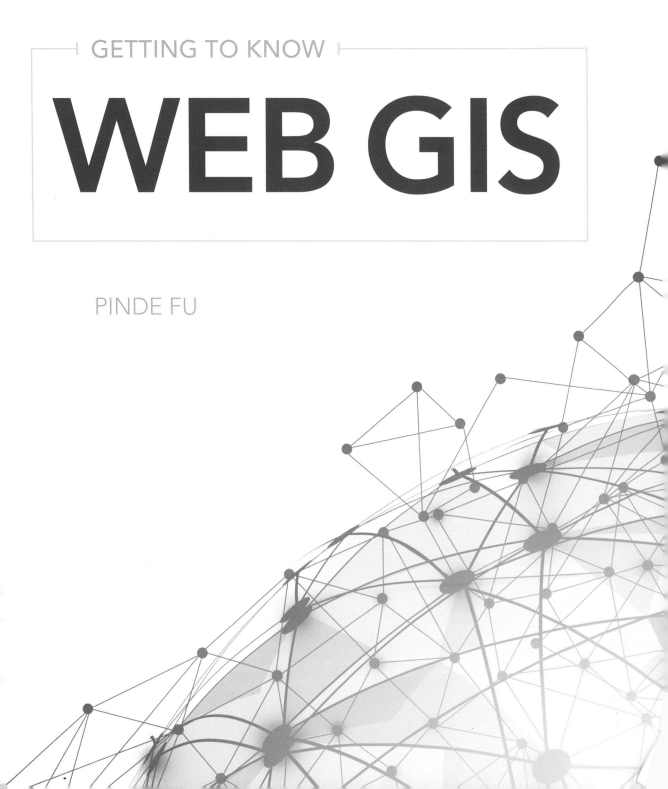

THIRD EDITION

GETTING TO KNOW

WEB GIS

PINDE FU

Esri Press, 380 New York Street, Redlands, California 92373-8100

Copyright © 2018 Esri

All rights reserved. First edition 2015. Second edition 2016

Printed in the United States of America

22 21 20 19 18   2 3 4 5 6 7 8 9 10

Library of Congress Cataloging-in-Publication Data

Names: Fu, Pinde, 1968- author.
Title: Getting to know Web GIS / Pinde Fu.
Description: Third edition. | Redlands, California : Esri Press, [2018]
Identifiers: LCCN 2018002808 | ISBN 9781589485211 (tp : alk. paper)
Subjects: LCSH: ArcGIS. | Geographic information systems. |
  Geography--Computer network resources.
Classification: LCC G70.212 .F79 2018 | DDC 910.285/53--dc23
LC record available at https://lccn.loc.gov/2018002808

Ask for Esri Press titles at your local bookstore or order by calling 800-447-9778, or shop online at esri.com/esripress. Outside the United States, contact your local Esri distributor or shop online at eurospanbookstore.com/esri.

Esri Press titles are distributed to the trade by the following:

In North America:
Ingram Publisher Services
Toll-free telephone: 800-648-3104
Toll-free fax: 800-838-1149
E-mail: customerservice@ingrampublisherservices.com

In the United Kingdom, Europe, Middle East and Africa, Asia, and Australia:
Eurospan Group
3 Henrietta Street
London WC2E 8LU
United Kingdom
Telephone: 44(0) 1767 604972
Fax: 44(0) 1767 601640
E-mail: eurospan@turpin-distribution.com

Contents

PREFACE

The third edition of *Getting to Know Web GIS* will help readers begin or continue their journey to learn about, understand, and apply Web GIS.

Web GIS is a promising field with great applicability to e-government, e-business, e-science, and daily life. The societal need for both the web and GIS has generated a strong and increasing demand for good Web GIS professionals. Professors need a lab book to teach Web GIS, and on-the-job professionals need a guide to teach themselves.

This book fits those needs. It teaches readers how to share resources online and build Web GIS apps easily and quickly. The book is a practical manual for classroom lab work and on-the-job training for GIS students, instructors, GIS analysts, managers, web developers, and a broad range of GIS professionals.

The first two editions made this book a best seller and a premier book in the Web GIS field. The new edition is the most current and comprehensive treatment of the Web GIS field, ranging from fundamental concepts to cutting-edge technologies. Universities around the world have adopted this book to teach such courses as Web GIS, Internet GIS, Online Mapping and Information Processing, Web GIS Principles and Applications, and Introduction to GIS.

The third edition builds on the success of the previous two editions to match newer releases of ArcGIS℠ Online and ArcGIS® Enterprise, with the following updates:

- A new chapter on image services and raster analysis
- Separate chapters with a more extensive experience using mobile GIS and real-time GIS
- Tutorials migrated from ArcMap™ to ArcGIS Pro®, a new and more powerful tool for processing and publishing GIS resources
- Newly added frontiers such as big data analysis, the Internet of Things (IoT), virtual reality, augmented reality, and artificial intelligence
- New products such as ArcGIS® Arcade, Survey123 for ArcGIS®, Workforce for ArcGIS®, Operations Dashboard for ArcGIS®, Drone2Map™ for ArcGIS®, Insights℠ for ArcGIS®, ArcGIS® API for Python, ArcGIS® GeoAnalytics Server, and ArcGIS® Image Server

In writing this book, I have tried to keep the following principles in mind:

- **Easy to apply:** You do not have to be a developer to build web apps. This book facilitates immediate productivity and teaches how to build engaging web apps without a single line of programming. Even the JavaScript programming chapter is relatively easy to follow, teaching readers how to adapt sample code.

- **Current:** Web GIS technologies advance rapidly. This book teaches state-of-the-art technical skills needed for building applications and managing projects.
- **Holistic:** Unlike books that focus on individual products, this book teaches Web GIS technologies as a holistic platform, from the server side to the browser, mobile, and desktop client side.

Each of the 10 chapters includes the following sections:

- A conceptual discussion that gives readers the big picture and the underlying principles
- System requirements that help instructors set up the lab
- A detailed tutorial with screen captures that confirm progress along the way
- Common questions and answers
- Assignments that allow readers to practice what they have learned
- Online resources with links to videos

This book is the result of the author's extensive working experience at Esri and teaching experience at Harvard University extension, University of Redlands, Henan University, and University of Texas, Dallas. This course and the set of labs have been well received by these universities and their students. Professors can use this book as the lab book for their Web GIS courses, and professionals at work can use this book for on-the-job training.

While retaining 10 chapters, the third edition has extended tutorials. Professors can break some of the longer chapters into multiple classes and choose which sections to teach based on available class time and software requirements. ArcGIS Online and a web browser are sufficient to complete most of the tutorials and sections, except for the following chapters:

- Chapter 5 requires ArcGIS Pro (2.0+) to publish tile layers, and optionally requires ArcGIS Enterprise (10.5+) to publish map image layers.
- Chapter 8 has optional sections that require ArcGIS Pro (2.1+) and ArcGIS Enterprise (10.6+) with ArcGIS GeoAnalytics Server to publish geoprocessing services and perform big data analysis.
- Chapter 9 has optional sections that require ArcGIS Pro (2.0+) and ArcGIS Enterprise (10.5+) to publish image layers.

Readers who don't have ArcGIS Online subscriptions can create an ArcGIS Online trial account, which also comes with authorization to use ArcGIS Pro. The sample data and teaching PPT slides for this book are available on the Esri Press book resource page at **esri.com/ gtkwebgis3**.

I welcome your feedback at **esripress@esri.com** and hope this book sparks your imagination and encourages creative uses of Web GIS.

ACKNOWLEDGMENTS

I would like to thank everyone at Esri Press who supported this book project. Special thanks go to Catherine Ortiz for inviting me to write this book, Stacy Krieg for managing the project, Mark Henry for his excellent editing that greatly improved the quality of the book, and the designer and other colleagues for their valuable contributions.

I would like to express my special appreciation to my Esri colleague Dr. Jie Chang, author of chapter 9. With his expertise on remote sensing, Dr. Chang's contribution—imagery services and raster analysis—adds an important component of Web GIS that the previous editions of this book did not have.

I am also grateful for the support of many other Esri colleagues and other contributors. This book would not be possible without the inspiration and support of Esri President Jack Dangermond. I sincerely thank Mourad Larif and Brian Cross for giving me the flexibility to work on the book; Clint Brown for providing guidance on the book contents; Derek Law and Geri Miller for reviewing the first edition; Professor Marci Meixler at Rutgers University for reviewing the second edition; and Jennifer Laws of Esri, Professor Jennifer Swift at the University of Southern California, Professor Bo Xu at California State University, and Danny Downing at the University of Florida for reviewing the third edition.

Thanks to Julia Guard for her assistance in designing the slides, Bethany Scott and Sarah Ambrose for providing figures on big data analysis, Al Pascual and Eric Wittner for providing materials on virtual reality and augmented reality, Morakot Pilouk for providing materials on real-time GIS and hosting the sample stream service for the real-time GIS chapter, and Javier Gutierrez and Matthew Miller for reviewing the 3D web scene chapter. Thanks to Nathan Shephard, Jeremy Bartley, Jeff Shaner, Weston Murch, Meg Hartel, Laurence Clinton, Jinnan Zhang, Zikang Zhou, Dawn Wright, David DiBiase, Jianxia Song, Ismael Chivite, Tif Pun, Derek Law, Wei-Ming Lin, Jinwu Ma, Rupert Essinger, RJ Sunderman, Maosour Raad, and Brenda Watson for sharing enlightened discussions on Web GIS and for their support.

This book was developed based on my work experience at Esri and my lecturing experience at Harvard University Extension, Henan University, the University of Redlands, the University of Texas at Dallas, the University of California, California State University, and many other universities. I want to thank my colleagues and students at these universities for providing feedback that has improved the content and structure of this book.

Finally, and most importantly, I would like to thank my family for their love and support.

FOREWORD

By Clint Brown, Esri

Much can be said about trends in information technology (IT). Perhaps first and foremost is the growing recognition and acceptance that GIS has evolved into an essential IT. GIS will be at the center of major advances in computing and IT. And many people are coming to the realization that geospatial expertise and systems will be essential for our planet's future.

Cloud computing—combined with IoT, smartphones, machine learning, and apps—enables an instrumented world where computing can be harnessed to analyze and respond to virtually any issue. GIS is already playing a critical role in these initiatives.

Web GIS provides a comprehensive approach for working with virtually all information sources. Further, the data in each individual organizational GIS is being brought together virtually to create a comprehensive GIS of the world in the cloud. Each of us is creating and maintaining our own layers; because all GIS layers register onto the earth, we are also contributing to and assembling a larger societal GIS for our planet—our individual GIS systems of record are being integrated, extended, and deployed as systems for insight as well as communal systems for engagement.

Virtually everyone is gaining access to this comprehensive, virtual GIS via web connections. GIS maps and apps are enabling all kinds of opportunities and engagements that extend far beyond our original system visions and goals.

In the past decade, GIS has expanded far beyond the professional GIS community. Some interesting developments helped to drive this expansion. As apps became popular, people everywhere began to use online maps—the foundation for shared GIS. Almost overnight, everyone began to recognize the power of GIS as an enabling information platform for improved understanding, decision-making, efficiency, and communication.

GIS provides a geospatial framework to integrate and interpret results. Over the past few decades, the mass adoption of the internet has led to a glut of information that we have come to know as "big data." GIS provides a geographic context to make sense of it all, while also providing the capability and context to analyze that data in real time.

People everywhere now recognize that GIS has become an essential computing infrastructure for every organization. Consequently, we expect the reach and impact of GIS to continue to expand at an accelerating pace.

CHAPTER 1

Web GIS introduction

This chapter introduces the concept of Web GIS with the ArcGIS Web GIS platform. The chapter begins with an overview of Web GIS and its advantages, introduces the ArcGIS Web GIS platform, lists the technical evolutions in Web GIS, explains the basic content types and user levels in ArcGIS Online and ArcGIS Enterprise, and demonstrates the workflow to build Web GIS apps using the Esri® Story Map Tour℠ template. This chapter familiarizes you with ArcGIS Online and ArcGIS Enterprise basic operations and workflows, and introduces flexible ways to build Web GIS apps that you will explore in other chapters.

Learning objectives
- *Grasp the concept and advantages of Web GIS.*
- *Understand ArcGIS Web GIS platform deploy models.*
- *Learn the components of the new-generation Web GIS platform.*
- *Understand the technical evolutions and trends in Web GIS.*
- *Learn the workflow for creating web apps.*
- *Work with GIS data in comma-separated value (CSV) files.*
- *Create and share web maps and web apps.*
- *Familiarize yourself with the Esri Story Map Tour template.*

What is Web GIS?

Web GIS is the combination of the web and GIS. The web removed the constraint of distance in cyberspace, and thus allows people the freedom to interact with GIS apps globally and access information almost instantly. Web GIS uses web technologies, including, but not limited to, Hypertext Transfer Protocol (HTTP), Hypertext Markup Language (HTML), Uniform Resource Locator (URL), JavaScript, Web Graphics Library (WebGL), WebSocket, and more.

The first operational GIS was developed in the 1960s by Roger Tomlinson. Since then, GIS has continuously evolved from a local file-based single computer system to a central database-based client/server system, often with multiple servers and many more client computers. The invention of the internet in the late 1950s and the World Wide Web in the early 1990s laid the foundation for an evolutionary leap toward Web GIS. In 1993, the Xerox Corporation's Palo Alto Research Center (PARC) developed a mapping web page, which marked the origin of Web GIS. In the 2000s, Web GIS evolved into a new generation—a system of distributed web services you can access in the cloud, as represented by the Esri ArcGIS platform.

Inheriting the power of the internet and the web, Web GIS offers many advantages:

- **Global reach:** you can share your geographic information easily within your organization and with people all over the world.
- **Large number of users:** you can share your app with dozens, or even millions, of users supported by the scalable cloud technology.
- **Low cost per user:** the cost of building one Web GIS app is often cheaper than building a stand-alone desktop solution and installing it for every user.
- **Better cross-platform capabilities:** web apps, especially those built with JavaScript, can run on desktop and mobile browsers running a wide range of operating systems, from Windows, Mac OS, and Linux to iOS, Android, and Windows Phone.
- **Easy to use:** Web GIS apps typically incorporate simplicity, intuition, and convenience into their design. Therefore, public users can use these apps without having prior knowledge.
- **Easy to maintain:** web clients can benefit from the latest program and data updates each time they access a web app. The web administrator does not have to update all the clients separately.

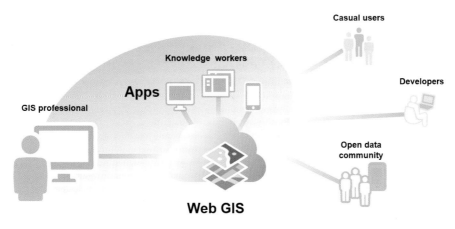

Web GIS presents a pattern for delivering GIS capabilities, and it enables all members of an organization to easily access and use geographic information within a collaborative environment. GIS professionals working on the desktop create and share information to the Web GIS and extend geospatial intelligence to broad users across organizations and throughout communities.

GIS is the science about locations, or The Science of Where™. The term has two meanings. One meaning is that GIS is itself a science, as the scientific basis for GIS technology. The other meaning is that GIS has been used for science as an effective tool for making scientific discoveries. The Science of Where is now recognized as a force for solving problems and understanding our world. Web GIS takes the science to a whole new scale, transforming how we share and collaborate, and revealing deeper insight into data. Web GIS unlocks and delivers the science to offices and homes and puts GIS technology in the hands of billions of people. Web GIS demonstrated immense value to government, business, science, and daily life. Recently, the concept and importance of spatial location has become more mainstream, and Web GIS awareness is growing more prominent in many organizations.

- **For government:** Web GIS offers an ideal channel for sharing public information services and delivering open data, an engaging medium for encouraging public participation, and a powerful framework for supporting decision making.
- **For business:** Web GIS helps create novel business models and reshape existing ones. It enhances the power of location-based advertising, business analysis, and volunteered geographic information, generating tremendous revenue, both directly and indirectly.
- **For science:** Web GIS creates new research areas and renews existing avenues of research.
- **In daily life:** Web GIS helps people decide where to eat, stay, and shop, and how to get from here to there.

ArcGIS is a Web GIS platform

Web GIS is central to Esri's strategic direction for implementing GIS as a platform. ArcGIS represents a cutting-edge and complete Web GIS platform that enables users to easily discover, use, make, and share maps from any device, anywhere, anytime.

ArcGIS is a new generation Web GIS platform that provides mapping, analysis, data management, and collaboration.

At the center of this Web GIS pattern is a portal, namely ArcGIS Online or Portal for ArcGIS, which represents a gateway for accessing all spatial products in an organization. The portal helps organize, secure, and facilitate access to geographic information products.

- Client applications on desktops, web apps, tablets, and smartphones interact with the portal to search, discover, and access maps and other spatial content.
- In the back-office infrastructure, the portal is powered by two components: GIS servers and ready-to-use content.

Web GIS deployment models

The ArcGIS Web GIS platform offers three deployment models.

- The **online model** uses only ArcGIS Online, the cloud-based Web GIS offering, in which all components are hosted in the public cloud. There is no hardware infrastructure for an organization to maintain because Esri manages and maintains ArcGIS Online.
- The **on-premises model** uses only ArcGIS Enterprise. Organizations manage the hardware and software infrastructure by themselves to implement the Web GIS platform. ArcGIS Enterprise includes four software components: Portal for ArcGIS, ArcGIS® Server, ArcGIS® Data Store, and ArcGIS® Web Adaptor. ArcGIS Enterprise has additional server roles such

as ArcGIS® GeoEvent™ Server, ArcGIS® GeoAnalytics™ Server, ArcGIS® Image Server, and ArcGIS® Business Analyst™ Server. Later chapters will have more details on these software components and server roles.

- The **hybrid model** combines parts of the online-based model with parts of the on-premises model. Hybrid deployment is by far the most common Web GIS deployment pattern. Details of such a model depend on an organization's business workflows and security requirements.

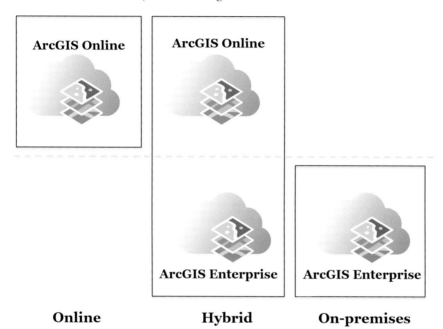

Online   **Hybrid**   **On-premises**

ArcGIS Web GIS platform can be deployed in three models. The dotted dash line represents the boundary between Esri-owned infrastructure and customer-owned infrastructure.

Compared to ArcGIS Enterprise, ArcGIS Online provides more ready-to-use contents, such as ArcGIS® Living Atlas of the World, and more ready-to-use-analysis services that are supported by these contents. ArcGIS Enterprise with its various license roles allows users to create more types of services than ArcGIS Online can create. But in general, ArcGIS Online and ArcGIS Enterprise share similar capabilities and similar workflows for creating services, web maps, and web apps. The tutorials in the book apply to both ArcGIS Online and ArcGIS Enterprise unless specifically stated.

Technology evolution and trends in Web GIS

Since its inception, Web GIS has been coevolving with geographic science and information technology. These evolutions and trends will be discussed in greater detail in later chapters.

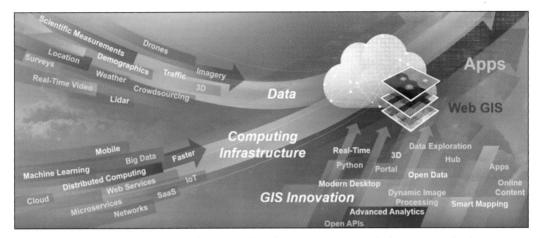

ArcGIS Web GIS platform coevolves with geographic information science and information technology.

Web GIS has exemplified the following evolution of stages and trends in technologies:

From closed websites to open geospatial web services

Early Web GIS applications were developed as independent websites. These websites were isolated from each other. It was difficult to share information and functions between them, and difficult to remix the websites to create new applications. In the later 1990s, web services technology was conceived. Web services can be thought of as building blocks that can be shared and remixed in versatile ways for building web applications. The GIS industry adopted the concept of web services in manufacturing Web GIS products. For example, ArcGIS Web GIS products fully support the web services architecture: server side provides ready-to-use services and allows users to create their own services; client side can consume and combine these services to create applications.

From one-way to two-way information flow

Early Web GIS products and applications mainly supported one-way information flow, which was from server side to client side. Users were merely the receivers of information. As time entered the new century, user-generated content (UGC) became a significant phenomenon and created a reversed information flow, from client side to server side. Volunteered geographic information (VGI) is the UGC of a geospatial nature and this was supported by Web GIS products. For example, ArcGIS facilitates VGI through editable feature services and hosted feature layers, mobile GIS apps, and browser-based apps. Users can view maps and perform queries, as well as conduct field surveys, collect data, and report events they saw. VGI provides unique values and perspectives to global observation, information sharing, and public engagement.

Portal technology is becoming essential

The word "portal" means gate or entrance. It was adopted in the mid-1990s to form new terms such as "web portals," referring to websites that serve as the gateway to other websites. Geoportals are gateways to geospatial information. They serve as the gateway or bridge between Web GIS servers and clients. Portals have become a core component of Web GIS technology. For example, ArcGIS Online and Portal for ArcGIS have geoportal capabilities. They facilitate the management, search, discovery, configuration, security, and remix of GIS data layers and services. Today, portals of different organizations can collaborate as hosts and guests, creating a "portal of portals" by using a distributed Web GIS pattern.

Cloud GIS accepted as the primary way to deliver GIS functions and ready-to-use contents

Cloud GIS, utilizing public and private cloud computing to provide GIS software and contents, has become the primary way to deliver GIS, not just functions, but also contents. Because of the vast contents and functions available from cloud GIS, the flexible pay-as-you-go or subscription-pricing model, and the reduced complexity and increased availability of services, cloud GIS, such as ArcGIS Online, has penetrated organizations that have not used GIS before or been able to afford GIS on their own.

Mobile is becoming the pervasive Web GIS client platform

As we entered the post-PC era, mobile devices have surpassed desktops and notebooks as the primary platform for accessing online information. Mobile devices are a part of people's life and work. "Mobile First" was one of the strategies of many industries, including the Web GIS industry. Vendors have given extra attention to mobile GIS. For example, Esri provides numerous mobile native apps and mobile-friendly browser apps to support people's and organization's need for mobile GIS. Mobile GIS is also associated with many frontiers in Web GIS, such as augmented reality (AR). AR can superimpose GIS data on top of a user's camera views and thus can augment a user's sense of reality. With the rapid advances in mobile GIS, the vision of using GIS for anything, anytime, anywhere, and by anyone is being realized faster than we can imagine.

Map visualization goes from 2D to 3D and virtual reality (VR)

With the increased client-side graphics processing power and the broader support of WebGL, Web GIS products such as ArcGIS Online and ArcGIS Enterprise can create and display thematic and photo-realistic 3D web scenes smoothly. Representing a big step from 2D online maps, 3D web scenes provide web users a more intuitive means to understand their data. 3D scenes are also important for indoor mapping, an ongoing trend in GIS. Even more intuitive than 3D, VR, such as ArcGIS 360 VR, allows users with certain visual wearables to immerse themselves into 3D city models by teleporting to static viewpoints and comparing different urban design scenarios. The immersive experience brings GIS data and geospatial understanding even closer to users.

1

2

3

4

5

6

7

8

9

10

Data source goes from static to real time and spatial temporal big data

Many elements in Web GIS are of a real-time nature, such as the incidents immediately reported by field crews or citizens using mobile devices, the concurrent measurements from sensor networks and smart cities. This massive amount of data presents challenges in real-time intake, processing, analysis, visualization, and storage. ArcGIS GeoEvent Server and the ArcGIS Trinity project utilize cluster computing and can ingest thousands and millions of sensor readings per second, process them, and store them in real time. Such products and research allow Web GIS to meet the requirements of the Sensor Web interface and the Internet of Things (IoT).

Web GIS becomes smarter and more intelligent

Map visualization is the first step toward data analysis. Online mapping becomes smart today. ArcGIS Smart Mapping can analyze the data automatically and suggest the best mapping style and the best defaults. This can help users, experts or novice, create beautiful and informative maps quickly. Web GIS goes far beyond mapping. ArcGIS GeoAnalytics Server can perform big data analysis using distributed computing, aggregate data in the context of both space and time, extrapolate new ideas from raw data, and bring superior intelligence to business decisions. More recently, Web GIS has started to use machine learning and artificial intelligence. For example, machine learning significantly improved the accuracy of online imagery classification. Artificial intelligence has been able to quickly identify the damaged locations from millions of facilities' photos, and ensure the damaged facilities are repaired early.

Paths to building Web GIS applications

The tutorials in this book teach readers how to build Web GIS apps. The ArcGIS suite of Web GIS products offers many paths to this goal.

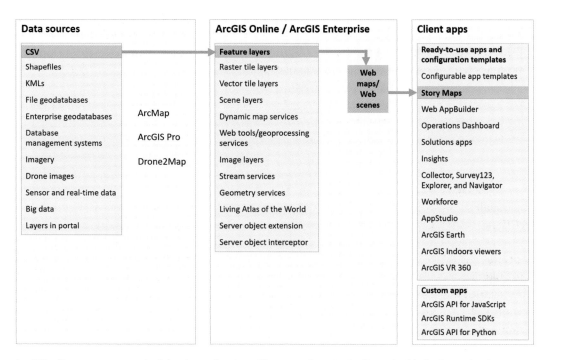

ArcGIS offers many ways to build web applications. The green lines in the figure highlight the technology presented in this chapter.

The previous figure presents the three basic tiers in Web GIS architecture and the generic workflow to build Web GIS apps:

- The data tier (on the left side of the figure) contains formats that range from simple CSVs managed with Microsoft Excel to sophisticated geodatabases managed with enterprise databases. This allows you to create map documents, toolboxes, and 3D scenes in ArcGIS® Desktop software, especially ArcGIS® Pro.
- In the middle tier of the figure, you can publish desktop resources to ArcGIS Online or ArcGIS Enterprise as several types of web layers and services. You can then add these layers and services to 2D web maps and 3D web scenes.
- Options for the presentation (or client) tier on the right side of the figure are essential apps. Apps range from ready-to-use ones that can be configured without programming to custom ones that use various web application programming interfaces (APIs) or software development kits (SDKs) to meet special requirements.

1

2

3

4

5

6

7

8

9

10

Start with ArcGIS Online

Cloud computing is based on the idea that many of the computing tasks that individual computers handle locally could operate more efficiently using huge computer centers connected through web technologies and provided as web-based services. Cloud GIS uses cloud computing technology to deliver GIS capabilities. This has helped users lower costs, reduce complexity, and quicken scalability.

ArcGIS Online (**www.arcgis.com**) is a cloud GIS. With ArcGIS Online, you can use and create web maps and scenes; access ready-to-use maps, layers, and analytics; publish data as web layers; collaborate and share maps; access maps from any device; and create apps from your maps. ArcGIS Online is a cloud GIS that provides the following services:

- **Infrastructure as a Service (IaaS):** you can upload your data and publish web layers to ArcGIS Online and host them on the ArcGIS Online infrastructure, which sits on top of Amazon EC2 (Elastic Compute Cloud) and Microsoft Azure. In this perspective, you would use the ArcGIS Online infrastructure, such as storage, CPU, and bandwidth.
- **Platform as a Service (PaaS):** you can build Web GIS apps without programming by using configurable apps or with programming by using ArcGIS® web APIs and ArcGIS® Runtime SDKs. In this perspective, you would use ArcGIS Online as a development platform for creating apps.
- **Software as a Service (SaaS):** you can use the rich collection of basemaps, thematic layers, analytical capabilities, and the countless and ever-increasing number of apps that are hosted in ArcGIS Online and published by Esri and its user communities. These capabilities are provided as a service from the cloud.

Adoption of ArcGIS Online and its quality of service

Before organizations add cloud GIS to their enterprise architecture, they first must assess the quality of services (QoS) of the cloud GIS. The following main factors represent QoS:

- **Performance:** How efficiently the system responds to user requests, usually measured in response time.
- **Scalability:** The ability to support a growing number of users without dramatically reducing performance.
- **Availability:** A measure of how often a system is accessible to end users, often measured in the percentage of time—for example, 99.99 percent.
- **Security:** The ability to provide confidentiality and secure access by authenticating the parties involved, encrypting messages, and providing access control.

ArcGIS Online provides reliable and trustworthy services in the four aspects listed. Based on many servers in the cloud and the use of high-performance computing technologies, ArcGIS Online hosts tens of millions of content items, millions of registered users, and responds to thousands of requests per second with fast performance, high scalability and availability. You can monitor ArcGIS Online availability in its health dashboard (**http://doc.arcgis.com/en/trust/ system-status**). ArcGIS Online follows a robust and effective framework to enforce security and protect user privacy. ArcGIS Online is certified as compliant with many federal and international security and privacy standards (see more information at **http://doc.arcgis.com/en/trust/ compliance/compliance-tab-intro.htm**). Because of the benefits of cloud computing and because of its high QoS level, ArcGIS Online has been quickly adopted by numerous government and commercial organizations around the world, from local to national governments as well as oil and gas, education, healthcare, law enforcement, banks, retailers, and more.

Web GIS information model

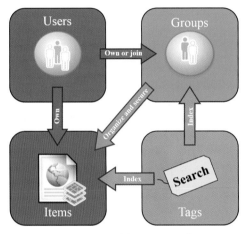

ArcGIS Online and Portal for ArcGIS information model.

The ArcGIS Online information sharing model has elements that include users, groups, content, and tags.

- Users can create and join groups.
- Users sign in to create and share content items, which can be a large variety of data, layers, and web maps and apps.
- Content items have tags. Tags are indexed so users can search and discover items more efficiently.
- Users can keep information to themselves, share with certain groups (not with individual users), share with their organizations, or share with everyone—the public. This allows other users to see and access the items. ArcGIS supports a variety of sharing levels.

Types of user accounts

ArcGIS Online and ArcGIS Enterprise support anonymous users, public users, and organizational users. Anonymous users can access the content shared with the public, if an organization has enabled anonymous access. Public users have limited abilities when creating and sharing content. Organizational accounts have levels, roles, and privileges.

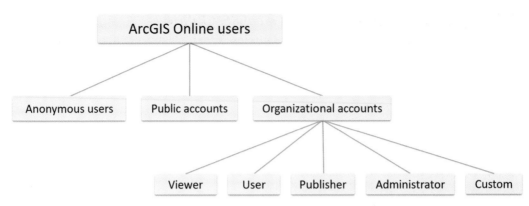

ArcGIS Web GIS user types.

There are two levels of organizational user accounts. Level 1 accounts are viewers only. Level 2 accounts can view, create, and share content. A role defines the set of privileges assigned to a member. ArcGIS defines a set of privileges for four default roles. Organizations may refine the default roles into a more fine-grained set of privileges by creating custom roles.

Privileges of anonymous users and default roles

Privileges	Anonymous users	Default roles			
		Viewer	User	Publisher	Admin
Use maps and apps shared with them.	✓	✓	✓	✓	✓
Use demographics, elevation analysis, geocoding, and network analysis; join groups without update capability.		✓	✓	✓	✓
Join groups with update capability; use subscriber content, spatial analysis, and GeoEnrichment; create content and groups; share maps, apps, and scenes; edit features.			✓	✓	✓
Publish hosted web layers; perform analysis.				✓	✓
Manage user accounts; manage organizational settings; create custom roles; set up enterprise log-ins; disable multifactor authentication on member accounts; manage credit budgets; view subscription status; create and own groups that allow members to update all items in the group.					✓

Main types of content items in ArcGIS Online

Five main types of content in ArcGIS Online relate closely to this book's main goal—learning how to build Web GIS apps.

Data
CSV, TXT, shapefile, GPX, geodatabase formats, and so on

Layers
Feature layers, tile layers, vector tile layers, scene layers, map image layers, image layers, stream service layers, KML, GeoRSS, WMS, WFS, WMTS, and so on

Tools
Query, geocoding, network analysis, geoprocessing services, and so on

Web maps

Web apps

The main types of content items in ArcGIS Online and Portal for ArcGIS.

Typically, a web app comprises one or more web maps, which in turn include or reference one or more layers.

- **Data:** ArcGIS Online supports data in a variety of formats, including CSV, TXT, Shapefile, GPX (GPS Exchange Format), and geodatabase.
- **Layers:** ArcGIS Online can host layers including the aforementioned data and can reference layers that include feature layers, tiled layers, vector tiles, map image layers, image layers, scene layers, CSV layers, tables, and Open Geospatial Consortium (OGC) standard layers such as GeoRSS, Keyhole Markup Language (KML), Web Map Service (WMS), Web Map Tile Service (WMTS), and Web Feature Service (WFS).
- **Web maps and scenes:** These maps interactively display geographic information that you can use to answer questions. A web map or scene (the 3D counterpart to a web map) comprises or references one or multiple layers.
- **Tools:** These tools perform analytical functions, such as geocoding, routing, generating PDFs, summarizing data, finding hot spots, and analyzing proximity.
- **Web apps:** Apps are the face of Web GIS and they are what brings Web GIS to life. Developers can program with ArcGIS web APIs to build web apps. However, you do not have to be a developer to create a web app. ArcGIS provides many templates that you can use to create impressive web apps without any programming.

Steps to creating Web GIS apps

Here is the typical workflow used to create web apps using ArcGIS Online or ArcGIS Enterprise:

1. Define the objectives of your information product.
2. Search for data layers in ArcGIS Online, ArcGIS Open Data (**http://opendata.arcgis.com**), your organization's ArcGIS Enterprise, and/or publish your data, maps, and toolboxes as web services or web layers.
3. Create and share your web map using the map viewer.
   - Add the layers you found and/or created to your web map.
   - Symbolize the layers (for some types of layers only) and configure pop-ups.
   - Save and share your web map.
4. Create and share your web app.

Browse the configurable apps to find a template that best suits your needs, and then use it to transform your web map into a web app. If no templates meet your requirements, use ArcGIS web APIs or Runtime SDKs to create your own app. After it is created, your app is private. You need to share it for others to search, discover, and use. There are different sharing levels.

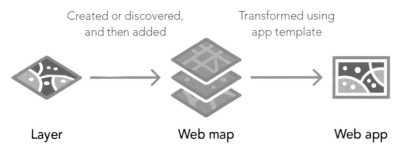

Created or discovered, and then added · Transformed using app template

Layer · Web map · Web app

ArcGIS Online and ArcGIS Enterprise allow users to easily create web maps by assembling various formats of layers and to create web apps from web maps by applying app templates.

Esri Story Map Tour

Story Map Tour is one of the most popular templates in ArcGIS Online. See two screen captures of the template in the figure and live samples at **http://arcg.is/2wOWLo2** (short for **http://storymaps.arcgis.com/en/gallery/#s=0&md=storymaps-apps:map-tour**).

Story Map Tour web app template works in a desktop browser (left), and on a smartphone.

The Story Map Tour template produces attractive, easy-to-use web apps that help you present geographic information with compelling photographic and video story elements. The template layout automatically rearranges itself to adapt to various screen sizes and can display a set of places on a map in a numbered sequence made for browsing. The template is designed for use in web browsers on the desktop, smartphones, and tablets.

Many scenarios benefit from this template:

- Show the world the work your government department, organization, or agency is doing or has done.
- Showcase key attractions of a city or region.
- Introduce a park and its features.
- Provide a tour of a campus, an outdoor art collection, or a historical district.
- Educate people about areas of scientific or geographic interest.
- Direct public attention to places you want to improve or protect.
- Create online photo or video journals of a trip or event.

This tutorial

In this tutorial, you will create a Web GIS app that introduces the main points of interest (POIs) in the City of Redlands, California.

Data: A CSV file contains data for the main POIs in Redlands, including longitude, latitude, names, descriptions, photo or video URLs, and thumbnail URLs.

The sample data for this entire book are available at **esri.com/gtkwebgis3**. Windows users can extract the files to C:\EsriPress. Mac users can create an EsriPress folder under your home directory and extract the files there. If you are in class, follow your instructor's instructions to download the data.

Requirements:
- Your web app should display a basemap (a street map or satellite imagery) of the city and POI locations, along with their descriptions and photos or videos.
- The web app is engaging and easy to use.
- The web app should work on desktops, tablets, and smartphones.

System requirements:
- Microsoft Excel or a text editor to create and edit your CSV data.
  - CSV easily represents points, though not complex geometric forms such as lines and polygons.
  - Excel automatically maintains correct CSV format (for example, adding correct quotes).
- A web browser.
- ArcGIS Online or Portal for ArcGIS.
  - A publisher or administrator level account: If your organization offers ArcGIS Online for Organizations or Portal for ArcGIS, ask your administrator or instructor to create an account for you. Otherwise, create a free trial account. Creating a trial organizational account will make you the administrator of the organization, which will allow you to create hosted feature layers.
- **Note to instructors:** Optionally, you can create a group for your students in which they can share their work with other members.

1.1 Create an ArcGIS Online trial account

You will skip this section if you already have an account for ArcGIS Online or Portal for ArcGIS.

If your organization has ArcGIS Online for Organizations or Portal for ArcGIS, please ask your administrator or instructor to create an account for you.

1. Open your web browser, navigate to **http://www.esri.com/arcgis/trial**.

2. Fill out the Sign Up for the ArcGIS Trial form:

    • Input your name, email, and other requested information.
    • Click Start Trial to submit the form. You will know the form has submitted correctly when a new page comes up that reads, "Confirmation email sent!"

Esri will send you a confirmation email for you to activate your account.

3. Check your email and click the activation URL link in the Activate Your Free ArcGIS Trial email.

4. On the activation page, fill in the fields, accept the terms and conditions, and click Create My Account.

Having created an ArcGIS Online for Organizations trial account, you have been made the administrator for your organization. You will be directed to the Set up Your Organization page.

5. On the Set up Your Organization page, fill in the fields. Then click Save and Continue. (Do not select Allow access to the organization through HTTPS only.)

Refer to the "Questions and answers" section for details on the HTTPS only option.

6. If prompted with the pop-up window asking if you want to download ArcGIS Pro and other software, click Continue with ArcGIS Online.

You will need ArcGIS Pro later in this book, but not now.

7. Click the Organization tab, click Editing Settings, click Security, and select Allow anonymous access to your organization's website.

This setting will allow your instructor to see your homework anonymously. Otherwise, your instructor will need to have an account in your trial organization to see your layers, maps, and apps.

8.   Click Save.

You have successfully created your trial organization account.

1.2 Create a hosted feature layer

Certain configurable apps require certain kinds of data content. The Story Map Tour template, for example, requires a layer of points. The layer should have the locations, captions, descriptions, photos or videos, and thumbnails associated with the tour points. You can organize your data in a CSV or point shapefile, feature service, map service, or other formats. This section will create a hosted feature layer for using in the web map and web app.

1.   If you have not already done so, navigate to **esri.com/gtkwebgis3**, or follow your instructor's directions to download the sample data for this book. Extract the files to C:\EsriPress on a PC or to EsriPress under your home directory on a Mac.

This tutorial works for both PCs and Macs, though the lab data directories differ on them. To simplify the tutorial instructions, the following tutorial mentions the data directory on the PC only.

2.   In Microsoft Excel, navigate to C:\EsriPress\GTKWebGIS\Chapter1\Locations .csv, and study its data format.

Name	Caption	Icon_color	Long	Lat	URL	Thumb_URL
Welcome to the City of Redlands	Located about 60 miles east of Los Angeles, replete with cultural, artistic and historical sites. Redlands, emerging as a regional leader, boasts small-town charm, and features a world center of geospatial information technologies. (\<i>\<a href="http://en.wikipedia.org/wiki/Redlands,_California" target="_blank">More info\</a>\</i>)	R	-117.182421	34.055448	https://googledrive.com/host/0BwLwen7VgLIFdW5uemtLeDJ0S3c/Cover.JPG	https://googledrive.com/host/0BwLwen7VgLIFMUk4UUZ0Q2d6YWM/cover.png
Esri	Headquartered in Redlands, Esri is a world leader in GIS software. Founded by Jack and Laura Dangermond in 1969, Esri now has 10 regional offices in the U.S. and a network of 80 international distributors, with about a million users in 200 countries. (\<i>\<a href="http://www.esri.com" target="_blank">Website\</a>\</i>)	R	-117.195688	34.056932	https://googledrive.com/host/0BwLwen7VgLIFdW5uemtLeDJ0S3c/Esri.JPG	https://googledrive.com/host/0BwLwen7VgLIFMUk4UUZ0Q2d6YWM/esri.png

This file has the main POIs in the City of Redlands. You will examine the sample data to familiarize yourself with the required fields. The first row of your spreadsheet provides the header. Below that, each row contains one tour point. For each point, the Story Map Tour template expects the following fields:

- **Name:** A short name identifying the point.
- **Caption:** A description of the point. Keep it short (less than 350 characters is recommended). The caption can include HTML tags to format the text or provide hyperlinks.
- **Icon_color (optional):** The color of each point. The valid values—R, G, B, and P—indicate red, green, blue, and purple, respectively.
- **Geographic Location:** You can describe geographic location by specifying longitude and latitude as Long and Lat (in decimal degrees), a single Address field containing a complete street address, or multiple fields (such as Address, City, State, and ZIP). This tutorial uses Long and Lat.
- **URL:** The full web address for the full-size image or video, starting with http://, https://, or //. The recommended image size is 1000 × 750 pixels, but other sizes will also work.
- **For videos:** The template does not include a generic video player. Instead, you will use the URL that a video hosting service, such as YouTube, provides for embedding videos via a link. You will append **#isVideo** to the end of the URL (for example, **http://www.youtube .com/embed/RM0eMdrPhEA#isVideo**). For YouTube videos, you will right-click the video being played, click Copy Embed Code, paste the code into Notepad, find the URL in the code, and append **#isVideo** to the end of the URL.
  - To use photos or videos on your computer, you must first upload them to some form of online storage, such as Flickr, YouTube, or your own web server.
  - If you have not yet collected your own images and videos, you can search for media through search engines and then copy their URLs. For images: Right-click an image. Select Copy Image Location in Firefox or Copy Image URL in Chrome. For Internet Explorer, select Properties and then copy the image address URL.

- **Thumb_URL:** The full web address of the thumbnail image (starting with http://, https://, or //). Images can fit to scale, but the recommended image size is 200 × 133 pixels.

3. Open a web browser, navigate to ArcGIS Online (**www.arcgis.com**) or your Portal for ArcGIS, and sign in.

| Home | Gallery | Map | Scene | Groups | Content | Organization | | GTKWebGIS ⌄ | Q |

You can familiarize yourself with the links at the top of the page:
- Home returns to the homepage.
- Gallery leads to featured maps and apps.
- Map goes to ArcGIS Map Viewer.
- Scene goes to ArcGIS Scene Viewer.
- Groups takes users to the My Groups page where you can create and join groups.
- Content links to the Content page, where you can see your own content, your groups' content, and your organization's content that are shared with you. Users with corresponding privileges can add and delete content items.
- Organization leads you to a page about your organization. If you are an administrator of your organization, the page includes tools you can use to manage your organization.
- In the upper-right corner of the page, the Search box and button allow you to search for content in the ArcGIS Online catalog.

4. Click Content.

5. Click Add Item ➕, and then click From my computer.

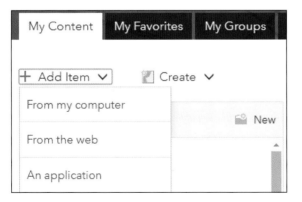

6.   In the Item from my computer window, perform the following tasks:

- For File, browse to C:\EsriPress\GTKWebGIS\Chapter1\Locations.csv, and click it. If you have published a file of the same name to your content before, rename your Locations.csv file to a unique name, and then select it.
- For Title, use the default, or specify a new one.
- For Tags, specify keywords, such as **Redlands Tour**, **GTKWebGIS**, and your organization name, as illustrated. Separate the keywords with commas.
- Make sure the check box next to Publish this file as a hosted layer remains selected.
- Leave the Use Latitude/Longitude option selected.
- Review the field types and location fields.
- Click Add Item.

Item from my computer ⑦

Add an item from your computer.

File:

| Choose File | Locations.csv |

Title:

| Locations |

Tags:

| Redlands Tour × | GTKWebGIS × |

Add tag(s)

☑ Publish this file as a hosted layer. (Adds a hosted layer item with the same name.)

Locate features using:
◉ Latitude/Longitude   ○ Address   ○ None, add as table

Review the field types and location fields. Click on a cell to change it.

Field Name	Field Type	Location Fields
Name	String	Not used
Caption	String	Not used

Time Zone: (UTC) Coordinated Universal Time ▼ ⑦

[Add Item] [Cancel]

Among the tags, GTKWebGIS stands for the title of the book. This tag indicates this item is based on the tutorial from this book.

The item details page appears as your CSV file is being published as a hosted feature layer. Each item in ArcGIS Online and Portal for ArcGIS includes an item page with a variety of information, actions, options, and settings organized by tab: Overview, Data, Visualization, Usage, and Settings. The tabs you see, as well as the options and information available on the tabs, depend on the item type, your privileges, and whether you are the item owner or administrator.

7.   Click each tab to familiarize yourself with the tabs on the item details page.

- The **Overview** tab includes overview information about an item, such as a description, tags, data source information, creation date, size, and sharing status. The tab also includes options to open, add to favorites, share the item, edit metadata, and add ratings and comments.
- The **Data** tab allows you to view, sort, and edit—if you are the item owner or the administrator—the attribute data of feature layers.
- The **Visualization** tab allows you to change default properties such as style, filter, pop-ups, and labels, of a feature layer without having to open the layer in map viewer.
- The **Usage** tab allows owners and administrators to see the usage statistics of the layer over time.
- The **Settings** tab allows editors and administrators to enable editing and configure other related settings for a feature layer.

8.   Click the Overview tab, click Share, share your layer with Everyone (public), and click OK.

This way, your web users can use your feature layer without having to log in.

In this section, you have created a feature layer, which will be used in the web map and web app you will be creating in the following sections.

1.3 Create a web map

You will make sure you are signed in before continuing with the remaining steps. Otherwise, you will not be able to save your web map and may lose your work. If you are not continuing from the last section, sign in to ArcGIS Online or your Portal for ArcGIS, go to your content, find and click the feature layer you created in the previous section.

1. Continuing from the last section, on the item details page of your feature layer, click Open in the Map Viewer drop-down list, and choose Add to new map.

This button has two options. You can Add to new map or Add to new map with full editing control. The latter allows owners and administrators to edit the layer data and schema. You don't need to edit the layer here. Either option will direct you to the map viewer. The locations of your data are displayed on the map automatically.

2. Familiarize yourself with the map viewer menu bar.

The ArcGIS Online map viewer helps users create, customize, and view web maps. On the menu bar, you will see the following buttons:

• The **Details** button toggles the panel on the left side of the web map canvas. This panel can display a web map's metadata, table of contents (TOC), or legend.
• The **Add** button is used to add a variety of layers into the web map.

- The **Basemap** button ▦ displays a gallery of underlying imagery that users can choose from.
- The **Analysis** button ⬚ leads to a rich set of analysis functions.
- The **Save** button 🖫 allows you to save your web map.
- The **Share** button ⇔ lets you select the people who will have access to your web map and choose how you will share it, either by embedding the web map in a webpage or by creating a web app from a template.
- The **Print** button 🖶 creates a hard copy of the current map view.
- The **Directions** button ◈ can calculate the best route from a starting location to the destinations you specify.
- The **Measure** button 📏 helps determine areas, distances, and a location's longitude and latitude.
- The **Bookmarks** button 📖 allows you to save a list of map areas so you can quickly select one and zoom to that map area.
- In the **Find address or place** text box, you can specify an address or place and find its location on the map viewer.

3.   If you are prompted with the Change Style pane, click Cancel.

The Map Tour template will use the style defined in the app template. There is no need to configure your layer style.

4.   In the Contents pane, point to the Locations Layer, click the More Options button •••, click Set Visibility Range, set the range to always visible (in other words, from World scale to Room scale).

5.   Zoom and pan the map to an extent that you will use as the initial extent of your web app.

6. On the menu bar, click the Save button 🖫 and choose Save.

7. In the Save Map window, enter the title, tags, and summary of your web map. Then click Save Map. Leave the web map open.

💡 **Tip:** For your homework, include your name in the title or tags, and include your university or organization name in the tags so that you and your instructor can easily identify your web map.

Save Map   ✕

Title:	Redlands City Tour (your name)
Tags:	Redlands Tour ✕ Your university name ✕
	GTKWebGIS ✕ Add tag(s)
Summary:	Main points of interest in the City of Redlands
Save in folder:	GTKWebGIS ▼

SAVE MAP   CANCEL

Congratulations! You have created a simple web map.

Typically, users need to configure pop-up windows and sometimes change styles on map layers. You will learn these skills in the next chapter. For this tutorial, the Map Tour template automatically handles the style of your layer, so you do not need to configure pop-ups or change layer style here.

1.4 Create a web app using a template

This step will transform your web map into a web app using the Story Map Tour template.

1. Continue from the previous section or sign in to ArcGIS Online or Portal for ArcGIS, and open the web map you just created. In the map viewer, click the Share button 🔗 on the menu bar, which opens the Share window.

2. In the Share window, select the check box next to Everyone (public) or next to the check box(es) next to the organization and group(s) with which you would like to share your web map.

Share   ×

Choose who can view this map.

Your map is currently shared with these people.

☑ Everyone (public)

☑ Esri Press - Official Book Site

☐ Members of these groups:

    ☐ OpenData_GTKWebGIS (Open Data)

Link to this map

http://arcg.is/1v0iWP

☑ Share current map extent

Embed this map

EMBED IN WEBSITE   CREATE A WEB APP

DONE

Note: Unless you share your web map and all the layers in the web map with everyone, a prompt will ask users to sign in whenever they open your web map and any web apps that use this map.

3. Click Create a Web App.

The Create a New Web App window opens presenting a gallery of the templates. If your organization has configured custom galleries, you may not see these same configurable apps as the figure.

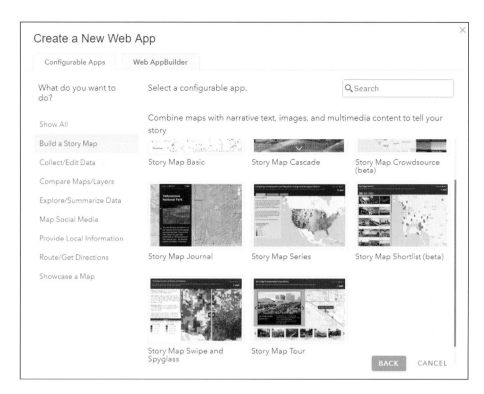

The templates are grouped and the groups are listed in alphabetical order. You can use the scroll bar to review the full gallery, or you can click a group name on the left to see the templates in this group.

4. Click Build a Story Map group on the left, find and click the Story Map Tour template, and then click Create Web App.

5. Leave the title, tags, and summary information as they are or make appropriate updates, and then click Done.

📖 **Note:** The check box, "Share this app in the same way as the map (Everyone)," is selected by default. With this check box selected, your web app will be shared with the public, along with your web map.

You have created your own informative and easy-to-use web app!

6. Click the Settings button, select the Three-panels layout, and click Apply.

7. Spend a few minutes exploring your new web app.

For example, you can navigate through the app's tour points by clicking the thumbnails, the arrows next to the photos, and the numeric icons on the map. If you click the thumbnail for the University of Redlands, for example, a video introducing the university appears.

Your web app is already created and saved. You will further configure it in the next section.

1.5 Configure your web app

Once you have determined that your app's tour points and their order, captions, and descriptions are correct, your app is complete. Optionally, you can enhance your application's features by using the template's builder mode. In this mode, you can add or import new tour points; update and delete existing images; set or update locations and descriptions; update the app title, subtitle, and logo; and change the app layout.

1.   If you are continuing from the previous section, go to step 3; otherwise, sign in to ArcGIS Online or your Portal for ArcGIS.

2. Click the Content tab, in your content list, find and click the web app you just created to go to its details page, and then click Configure App.

3. Familiarize yourself with the builder mode.

- The **pencil** icon  means that you can update nearby text, such as titles, subtitles, image captions, and descriptions.
- The **Change media** and **Change thumbnail** buttons can be used to change the URLs of media and thumbnail locations.
- The **Add**, **Organize**, and **Import** buttons allow you to interactively add more locations, change the order of points, and import media from Flickr, Picasa, Facebook, YouTube, or a CSV file.

Now, you will change the Esri photo into a video.

4. Click the Esri thumbnail image. Click Change media, then click Video. Remove the current URL, enter **https://www.youtube.com/embed/ RM0eMdrPhEA**, and click Apply.

The video loads into the picture frame. Refer to the "Questions and answers" section to learn how to get video URLs for use with the Map Tour template.

Next, you will change the thumbnail for Esri to a new one that indicates a video.

5. Click Change thumbnail. Replace the current URL with **https://bit.ly/2J4o2FI**, and then click Apply.

6. Click Save to save your changes.

STORY CONFIGURATION

| SAVE | SHARE | SETTINGS | HELP |

1 unsaved change : Tour is shared publicly

In the following steps, you should save your work regularly to prevent losing your changes.

7. Above the thumbnail carousel, click Organize.

| Add | Organize | Import |

The **Organize the tour** window allows you to delete tour points and drag pictures to change their order.

8. At the bottom of the Organize the tour window, select the check box for Use the first point as introduction (does not appear in carousel). Click Apply to close the Organize the tour window.

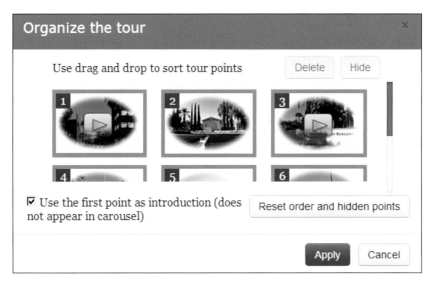

Selecting the check box sets the first record in your CSV as the introductory image to your app. This selection allows you to start your tour by showing a compelling image and an introductory caption to set the scene. The location of this record will not be shown on the map as a numbered point in your tour.

Optionally, you can click the Import button next to the Organize button to import tour points and media from Flickr, Picasa, YouTube, or an additional CSV file. You can also add additional tour points manually by clicking the Add button and filling in the media, name, and location information.

9. In the page header, click Settings.

Clicking Settings opens the Story settings window with the following tabs:
- **Layout:** Choose between the Side-panel Layout, Three-panel Layout, and Integrated Layout.
- **Colors:** Choose from predefined color themes or define your own theme.
- **Header:** Set the header logo and share links.
- **Data:** No configuration is needed here. The sample CSV you use has all the fields named properly for the map tour template to use.
- **Extent:** Define the initial map extent that users will see when the app first opens.
- **Zoom Level:** Specify a scale to which the map will automatically zoom whenever the app user goes from one tour point to another (but if users manually zoom in or out, the map tour app respects their choice and no longer applies your auto zoom level).

10. Click the Zoom Level tab and set the Scale/level to 1:5K (level 17) as illustrated.

This scale allows users to see the selected POI and its adjacent area.

Story settings   ✕

| Layout | Colors | Header | Data | Extent | Zoom Level |

Set zoom for story points following introduction (optional).

Scale/level: 1:5K (level 17) ▾

Apply   Cancel

11. Click the Header tab, change the logo and text if needed, and then click Apply.

For example, you can add your name to the header so that your instructor can easily tell who created your application. Optionally, you can also exchange the logo for your organization's logo.

Story settings   ✕

| Layout | Colors | Header | Data | Extent | Zoom Level |

Customize the header logo (maximum is 250 x 50px).

● Esri logo   🌐 esri
○ No logo
○ Custom logo

    Image URL

    Click-through link

Customize the header top right link.

Text: by Peter   ☑ Facebook
                                              ☑ Twitter
Link: http://storymaps.arcgis.com   ☑ Share

Apply   Cancel

Examine the application to see if there is anything else you would like to configure. If so, you can make and apply further changes.

12. In the page header, click Save to save your work.

1.6 Share your web app

In step 5 of section 1.4, you created and shared this web app with the same people with whom you shared your web map. Now you will share the URL of this web app with your audience so that they can see your web app.

1. Click Share in the page header. If you see a message saying your tour is not shared, share your tour publicly.

2. In the Share your Tour window, click Open to preview your web app.

3. Share the tour URL with your audience (for example, by copying and sending the URL through email, or by displaying the URL link on your organization's home page).

4. Test your web app on smart mobile devices.

Open your app in your smart device's browser. To do this easily, send the URL to yourself, check your email on your smart device, find the message, and then click the URL.

Configurable apps use responsive web design technologies and can change their layouts to best fit various devices with different screen sizes. You will find they work well on iOS, Android, and Windows Phone tablets and phones.

In this tutorial, you have created a user-friendly, informative, and cross-platform web app. The app meets all the requirements listed earlier in this chapter—it displays a basemap and POI locations, their descriptions, and any photos or videos; is engaging and easy to use; and works on desktops, tablets, and smartphones.

You can create a map tour app using other workflows. In addition to pictures and videos, you can also display webpages and 3D web scenes. See the Resources section for more information.

1

QUESTIONS AND ANSWERS

1. How can I get the latitude and longitude coordinates of a location or address?

   Answer: You can use ArcGIS Online or Portal for ArcGIS map viewer to do so. If you know where the location is, navigate there directly on the map. If not, you can type the address in the map viewer Search text box, and click the Search button. After the address is found and the map is centered to the location, click the zoom in button until you can no longer zoom in. Next, click the Measure button ⬚, click the Location button ⬚, and then click the location on the map. The location's longitude and latitude display under Measurement Result.

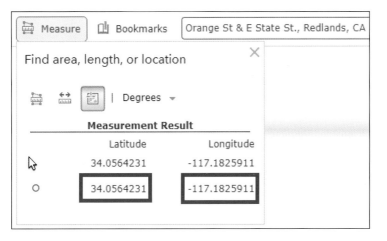

2. Locating longitudes and latitudes manually (one by one) can be slow work. Is there a more efficient way to define the locations of my points?

   Answer: Use addresses, feature classes, or geotagged media if you have them. In your CSV, specify the addresses of your points in one or multiple address fields such as Address, City, State, and ZIP. When you create your feature layer, ArcGIS Online and Portal for ArcGIS will geocode these addresses and find their locations automatically.

   If you have your tour points in a shapefile or file geodatabase, you do not need to create a CSV file. You can zip your shapefile (do not include a folder) or file geodatabase (include the folder of the file geodatabase), and then create your feature layer using this zip file.

   If you have geotagged media, for example, photos taken using your smartphone with location enabled for your camera, you can create an empty web map and import these photos using the app builder mode.

3. After publishing my CSV to ArcGIS Online as a feature layer, I need to update my CSV. Will the changes to my CSV automatically update in my feature layer, web map, and web app?

Answer: No, but you can overwrite the feature layer, or edit your data directly in ArcGIS Online or Portal for ArcGIS. Once your CSV data has been published to ArcGIS Online, it is uploaded to the cloud. Your feature layer, web map, and web app will use this cloud copy, rather than your local copy. To use your new CSV data, go to the details page of your feature layer, click the Update Date button, select Overwrite Entire Layer to re-create your feature layer using the new CSV. This will preserve your feature layer id, and thus won't break the web maps and web apps using your layer.

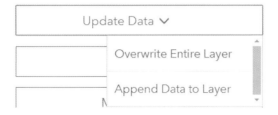

You can also edit data directly in ArcGIS Online or Portal for ArcGIS. You can go to the details page of your feature layer, click the Data tab, and double-click a value in the table to change it. Or you can click the Overview tab, click Open in Map Viewer, select Add layer to new map with full editing control, click the Edit button in the map viewer, click a tour point on the map, and click Edit in the pop-up. Then you can edit the attribute values and move the point on the map.

4. I submitted my homework URL, but my professors can't access it. Why? How can I fix this problem?

Answer: First, make sure your layer(s), web map, and web app for this assignment are all shared with the public. Next, check if your organizational settings allow anonymous access. Refer to step 7 in section 1.1. If you still have the problem, go to **http://storymaps.arcgis.com/en/my-stories/**, sign in, and click the Check Stories button to find the causes.

5. Should I enable my ArcGIS Online organization to use HTTPS only?

Answer: It depends. In general, it is recommended that you enable HTTPS only, which gives you enhanced security. If you enable HTTPS only, this implies that all the services and layers that your web maps and apps will be using must be configured to support HTTPS because web browsers do not allow the mixing of HTTP and HTTPS content. Some organizations still have map services and geoprocessing services running on the ArcGIS Server that doesn't have HTTPS configured. These services can be loaded via HTTPS. If you will use such services in your web maps and apps, you should disable the HTTPS only option (refer to step 5 in section 1.1.).

6. I need to upload my photos to a web server to get their URLs. Is there a web hosting service that you would recommend?

   Answer: There are many ways to upload and host your photos. Here are a couple easy ways:
   - Upload your photos to **https://imgur.com**, and then find the direct links to each photo.
   - Upload your photos to ArcGIS Online. Refer to **https://blogs.esri.com/esri/arcgis/2018/01/05/photos-images**.
   - Upload your photos to Flickr, and then import them to your map tour app.

7. I can drag and drop my CSV to the map viewer directly. How does this layer differ from a hosted feature layer?

   Answer: The former can be used for quick demos of small amounts of data, but it is not recommended in general. The latter has better reusability, supports larger data sizes, and supports more operations.

	Embedded layers (layers added to web maps directly)	Hosted feature layer and feature service
Storage	The data is stored or hardcoded in the web map.	The data is stored in its own layer. Web maps just store a reference to this layer.
Data size limitation	Limited. Up to 1,000 features per layer, or 250 features per layer if using geocoding.	1G per hosted feature layer; unlimited for feature services (requires ArcGIS Enterprise).
Reusability	Not reusable. To use the data in a different web map, you would have to add the CSV to the web map one more time. This essentially creates another copy.	Reusable. You can reference a feature layer—one copy of truth—in as many web maps and apps as you want.
Capabilities	Doesn't support query, search, and charting in most ArcGIS client apps such as Web AppBuilder for ArcGIS.	Support query, search, and charting in most ArcGIS client apps such as Web AppBuilder for ArcGIS.

8. It's slow to upload my photos to a web server and find the latitudes/longitudes of these photos manually. Are there ways to create Story Map Tour apps without me having to upload photos and locating them myself?

Answer: This tutorial teaches the generic workflow from data > layer > web map > web app. This workflow is applicable when you create web apps using other templates or app builders.

   You can create map tour apps using many other ways, some of which are easier though they are specific to the map tour template. For example, you can use Story Map Tour Builder, and choose to upload your images. This approach will upload your photos for you. If your photos are geotagged, the location of each photo is automatically extracted. Refer to **http://storymaps.arcgis.com/en/app-list/map-tour/tutorial** for more details.

ASSIGNMENT

Assignment 1: Choose from the following topics, and create an app using Esri Story Map Tour to showcase your chosen topic:

- Your personal story (where you were born, where you moved, where you went to school or worked, and so on)
- Your city's key attractions
- The landmarks, buildings, and departments on your campus
- Places you have visited in the past or during a recent vacation
- Branches of a bank or supermarket in your city or region
- Projects that your organization has accomplished or is working on
- Locations of key environmental interest (for example, largest/oldest trees) or historic interest (for example, oldest houses)
- Other interests

What to submit:

- Your web app URL

Resources

"Add YouTube videos to your Story Map Tour," Bern Szukalski, http://blogs.esri.com/esri/arcgis/
2015/08/12/add-youtube-videos-story-map-tour.

"Adding links to captions in your Story Map Tour," Bern Szukalski, http://blogs.esri.com/esri/
arcgis/2015/08/12/add-links-map-tour.

"ArcGIS Online: Administration Basics," https://www.esri.com/training/catalog/580fc524a4a46d1
72b116873/arcgis-online:-administration-basics (or http://arcg.is/2Duh8aA).

"ArcGIS Online: Getting Started," https://www.youtube.com/playlist?list=PLGZUzt4E4O2IJt1O_
OTDFR-3dUpiCZGKf (or http://bit.ly/2cy3IM3).

"ArcGIS Online Help," http://doc.arcgis.com/en/arcgis-online.

"ArcGIS Online: Sharing Basics," http://www.esri.com/videos/watch?videoid=gQ68gWcN1Mk&channelid
=UCgGDPs8cte-VLJbgpaK4GPw&title=arcgis-online:-sharing-basics (or http://arcg.is/2mtn6QO).

"Chapter 1, GIS Provides a Common Visual Language," *The ArcGIS Book*, Christian Harder and Clint
Brown, http://learn.arcgis.com/en/arcgis-book/chapter1.

"Create a Story Map Tour from a Google Sheet," Owen Evans, http://blogs.esri.com/esri/
arcgis/2015/08/12/create-a-map-tour-from-a-google-sheet.

"Get Started with ArcGIS Online (2017)," https://www.esri.com/training/catalog/57eb2e47ee85c0f5204be
b1d/get-started-with-arcgis-online-(2017)/ (or http://arcg.is/2AZsssh).

"Get Started with Story Maps," http://learn.arcgis.com/en/projects/get-started-with-story-maps.

"GIS—The Science of Where," Michael Goodchild, https://www.youtube.com/watch?v=iyBOsaFLTrU (or
http://bit.ly/2FCxnTp).

"Make a GeoPortfolio," http://learn.arcgis.com/en/projects/make-a-geoportfolio.

"Make a Map Tour Story Map," Rupert Essinger, http://www.esri.com/esri-news/arcwatch/0513/
make-a-map-tour-story-map.

"Set Up an ArcGIS Organization," http://learn.arcgis.com/en/projects/set-up-an-arcgis-organization.

"Story Map Tour," https://storymaps.arcgis.com/en/app-list/map-tour.

"Story Map Tour source code," https://github.com/Esri/storymap-tour.

CHAPTER 2

Hosted feature layers and Esri Story Maps

Apps are important because they are the face of Web GIS: They bring Web GIS to life. You created a web app in chapter 1. The ArcGIS Web GIS platform offers far more web apps. This chapter presents an overview of ArcGIS configurable apps, which include apps from ArcGIS Online, Portal for ArcGIS, Esri Story Maps, and ArcGIS Solutions. The chapter then introduces the basic components in today's Web GIS apps and the foundation supporting these components, which is web services technology. You'll learn about feature services and feature layers, which are the most common types of operational layers. The chapter then teaches how to create a hosted feature layer, design layer styles using smart mapping, and configure layer pop-ups with ArcGIS Arcade. The tutorial section further illustrates how to use layers from the Living Atlas of the World, and how to create a web app using Story Map Journal.

Learning objectives

- *Understand the suite of ArcGIS configurable apps.*
- *Understand the suite of Story Maps templates.*
- *Grasp the concept of web services.*
- *Create hosted feature layers using geocoding.*
- *Design layer style using smart mapping.*
- *Configure pop-ups with multimedia and Arcade.*
- *Explore Living Atlas of the World contents.*
- *Create an app using Story Map Journal.*

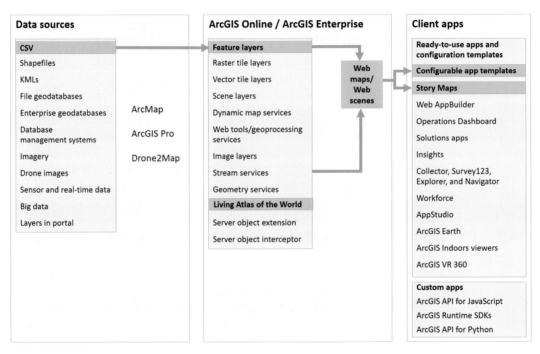

ArcGIS offers many ways to build web apps. The green lines in the figure highlight the technology presented in this chapter.

Apps, the face of Web GIS

GIS apps are lightweight map-centric computer programs that end users use on mobile devices, in web browsers, or on desktops. Web GIS end users directly interact with apps. The back-end data, services, and server computation power of Web GIS become live and useful through GIS apps. ArcGIS provides ready-to-use web apps, including configurable apps, mobile apps, app builders, and more. Mobile apps and app builders such as Web AppBuilder for ArcGIS will be introduced in other chapters. This chapter focuses on configurable apps.

ArcGIS configurable apps enable the broad user community to build engaging apps with no GIS or web development skills. ArcGIS platform provides the following configurable apps:

- **ArcGIS Online and Portal for ArcGIS app templates:** These apps are also called configurable apps.
- **Esri Story Maps:** Some of the Story Maps apps have become ArcGIS Online and Portal for ArcGIS app templates.
- **ArcGIS Solutions apps:** These apps provide data and map and app templates for you to quickly jumpstart your applications projects.

Using ArcGIS configurable apps generally takes the following three steps: choose, configure, and deploy.

Using ArcGIS configurable apps generally takes three steps: choose, configure, and deploy the app.

- **Choose:** In this step, you discover the data, maps, and configurable apps that match your app requirements. Consider the following factors:
  - **Purpose:** Who is your intended audience? Where and how will your audience use your app? What key points do you want to communicate?
  - **Functional requirements:** What critical functionalities support your purpose?
  - **Aesthetic:** How will the app's layout and color scheme support your brand or message?
- **Configure:** Configure the apps to use your data, and brand the apps for your organization.
  - ArcGIS Online and Portal for ArcGIS apps templates typically have a configuration user interface.
  - Esri Story Maps provide a builder user experience.
  - Solutions apps provide a configuration file that often requires manual editing.
- **Deploy:** Deploy your new apps for your end users.
  - Esri automatically hosts apps created using ArcGIS Online and Portal for ArcGIS app templates and Story Maps in the cloud. If needed, you can download the source code for these open-source apps and host the apps on your own web servers.
  - Solutions templates need to be deployed to and hosted on your own web servers.

ArcGIS Online and Portal for ArcGIS app templates

ArcGIS Online and Portal for ArcGIS provide many app templates. To help you choose the right one, ArcGIS organizes these apps based on their purposes in the following categories:

- **Build a story map:** These apps are adopted from Esri Story Maps.
- **Collect and edit data:** These apps primarily collect data. These apps fall into the subcategories of crowdsourcing and general editing.
- **Compare maps and layers:** These apps are focused on comparing geographic phenomena.
- **Display a scene (3D):** These apps allow you to interact with 3D scenes outside of scene viewer.
- **Explore and summarize data:** These apps allow your users to interact with attributes and in some cases other services to facilitate a deeper exploration of the content in your web map.
- **Make a gallery:** These apps create a gallery of maps, apps, or other content that can be used as a convenient access point for all your geographic content. These apps require a group.
- **Map social media:** These apps include social media content to supplement your message with relevant content.
- **Provide local information:** These apps highlight the resources available at a location. Options include highlighting all features within a certain distance of a location and communicating that a user's address is located within a certain geographic area.
- **Route and get directions:** Use these apps to provide driving directions from a user-defined starting point to the geographic features within your map.
- **Showcase a map:** These apps include many options for presenting thematic or general maps. The apps include legends, descriptions, and other basic tools to assist users in understanding the message of the map.

Esri Story Maps

Every story happens somewhere. Web GIS can enhance storytelling by visually and intuitively illustrating the "where" component of every story. Esri Story Maps (**https://storymaps.arcgis.com**) are simple web apps that combine interactive maps, multimedia content—text, photos, video, audio, and intuitive user experiences—to tell stories about the world. Story maps use geography to organize and present information.

Maps + **Story** + **Multimedia** = **Story Map**

Esri Story Maps combine interactive maps, multimedia content, and user experiences to tell stories.

Main types of configurable app templates provided by Esri Story Maps

Presenting sequential, place-enabled photos or videos		
Story Map Tour		Presents a set of photos or videos along with captions, linked to an interactive map
Presenting a rich multimedia narrative		
Story Map Journal		Creates an in-depth narrative organized into sections presented in a scrolling side panel, including media types such as a map, 3D scene, image, and video
Story Map Cascade℠		Creates a visually and editorially engaging full-screen scrolling experience blending narrative text, maps, 3D scenes, images, and videos
Presenting a series of maps		
Story Map Series℠ - Tabbed Layout		Uses a set of tabs
Story Map Series - Side Accordion Layout		Uses an expandable panel
Story Map Series - Bulleted Layout		Uses numbered bullets, one per map

Main types of configurable app templates provided by Esri Story Maps

Presenting a dynamic collection of crowdsourced photos		
Story Map Crowdsource℠		Publishes and manages a crowdsourced story to which anyone can contribute photos with captions. Display the collected photos along with their locations, captions, and descriptions.
A curated list of points of interest		
Story Map Shortlist℠		Presents a set of places organized into a set of tabs based on themes.
Comparing two maps or two layers of a single web map		
Story Map Swipe℠		Users can slide the swipe tool back and forth to compare one map theme to a second map theme.
Story Map Spyglass℠		Similar to Swipe but enables users to peer through one map to another using a spyglass function.
Presenting one map		
Story Map Basic		Presents a map via a very simple minimalist user interface.

You can create a story map in either ArcGIS Online or Esri Story Maps:

- **ArcGIS Online:** For example, you can share a web map, click Create a Web App, and choose a story map template.
- **Esri Story Map website:** You can select a type of story map and follow the builder wizard.

ArcGIS Solutions apps

ArcGIS Solutions provides a gallery of free templates to jumpstart your projects. The templates typically include app source code. Some templates also include data models, layers, maps, and sample web services. The data model and map styles are created based on industry best practices and emerging trends. ArcGIS Solutions apps cover almost all industries. The apps focus mostly on the web and mobile platforms and occasionally on the desktop platform.

You can search for templates in ArcGIS Solutions by products, industries, and keywords. For each template, you can read its introduction, requirements, and contents in the package. You can also try the app live, download the template, configure the app, and deploy it.

ArcGIS Solutions provides a rich collection of configurable apps for almost all industries.

Configuring ArcGIS Solutions web apps typically involves editing the config.js file. In this file, you replace the default URLs of the maps and layers with those URLs of your maps and layers and replace the default attribute field names with those field names of your data. Deploying these web apps typically requires a web server such as Apache, Microsoft IIS (Internet Information Service), or other products. You can set up your own web server or purchase a web hosting service. Once you have a web server, deploying a JavaScript web app is essentially copying the files to a folder under your web server's web root, for example, C:\inetpub\wwwroot for IIS.

Basic components of a Web GIS app

Today's Web GIS best practices recommend that a Web GIS app should have basemaps, operational layers, and tools. ArcGIS supports this practice and makes it easy for you to create Web GIS apps.

Web GIS app   **Basemaps**   **Operational layers**   **Tools**

The basic components of a Web GIS app.

Basemap layers

Basemaps provide a reference or context for your Web GIS app. ArcGIS provides a collection of fast-responding basemaps. Most of the time, you can use them without worrying about creating them.

- ArcGIS provides a gallery of basemaps, including image tiles and vector tiles. These maps all have multiple scales with global coverage.
- In addition, you can use your own map services as basemaps (open the map viewer, click Add > Add Layer from Web, and select the Use as Basemap box).
- ArcGIS provides a global elevation service on which basemaps can be draped. This service supports 3D web scenes.

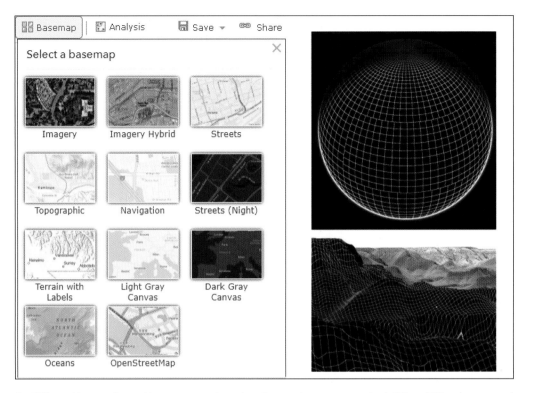

ArcGIS provides a gallery of basemaps and an elevation service to support both 2D and 3D web maps and apps.

Operational layers

Operational layers are theme layers that you and your end users mostly use.

- You can use layers from ArcGIS Online, Living Atlas of the World, and ArcGIS Open Data (**http://opendata.arcgis.com**) as your operational layers. These layers span a range of subjects and can support maps and apps of almost every subject. You can search in these rich collections and discover layers that fit your needs.
- You can also use your own data as operational layers. ArcGIS Online and ArcGIS Enterprise support operational layers from a variety of formats including CSV, TXT, GPX (GPS exchange), shapefiles, and a variety of web services types.

Tools

Tools perform tasks beyond mapping, including common tasks such as query, geocoding, routing, and more specialized tasks such as the workflows and big data analysis that implement specific logic for an enterprise.

- Most ArcGIS web layers support flexible queries.
- ArcGIS Online provides rich and extensive spatial analysis tools for you to ask questions and solve spatial problems.
- ArcGIS Enterprise provides standard analysis tools, big data analysis tools, raster analysis tools, and allows you to publish custom web tools, in other words, geoprocessing services.

Web services, the foundation of today's Web GIS

Web services technology is at the heart of today's Web GIS. The previously mentioned three types of components of Web GIS apps are almost all based on web services. For example, basemaps are often tile services, operational layers are often feature services, and many tools are based on geo-processing services.

Early Web GIS was not based on web services; instead, it consisted mostly of stand-alone websites where GIS data and functions were available only to their own clients and couldn't be reused in other systems. This situation greatly limited the reusability of Web GIS resources. As a result, the level of effort to create Web GIS apps was high. In the later 1990s, web services technology was conceived. A web service is essentially a program that runs on a web server and exposes programing interfaces for clients to consume over the web. Web services have many advantages, especially their flexibility in being reused and remixed in many web apps.

Over the years, the main web services interface style has evolved from Simple Object Access Protocol (SOAP) to Representational State Transfer (REST). REST is considered "the command line of the web," and relies on URLs to send requests and pass parameters. REST became the preferred web service interface type because of its simplicity and efficiency for the web environment. With REST, every resource has a URL, also called endpoint, and can be accessed via this URL. Therefore, you can add ArcGIS web services to the ArcGIS Online map viewer via their URLs. You will learn more about REST in other chapters later in this book.

The GIS industry quickly adopted web services technology and reformed Web GIS products based on web services architecture in the early 2000s. Today's Web GIS products are designed to support the publication, discovery, and use of GIS web services. A GIS service represents a GIS resource—such as a map, locator, or toolbox—that is located on the server and is made available to client applications. For example, ArcGIS Online provides collections of ready-to-use GIS services. ArcGIS Online and ArcGIS Enterprise allow publishers to publish many types of GIS services. ArcGIS Online and Portal for ArcGIS allow users to discover and use GIS services in web maps, web apps, and mobile apps. This chapter focuses on feature services or feature layers. You will learn about other types of services in other chapters of this book.

Feature services and hosted feature layer

Feature services and **hosted feature layers** are the most commonly used layer type for operational layers. A feature service in ArcGIS often refers to a feature service published to ArcGIS for Server. A hosted feature layer refers to a feature service published to ArcGIS Online or Portal for ArcGIS. "Layer" here refers to a content item in ArcGIS Online and Portal for ArcGIS. "Hosted" refers to the fact that they are hosted in the ArcGIS Online cloud or the underlying data is stored in the ArcGIS Data Store. Feature services and hosted feature layers can support read and/or write data access. This chapter will focus on the read-only hosted feature layers. You will learn about the writable feature layers in the mobile GIS chapter.

You can publish hosted feature layers to ArcGIS Online or Portal for ArcGIS from ArcGIS Pro, or simply a web browser.

- **Create a feature layer from your own data** (for example, CSV, shapefiles, GeoJSON, and file geodatabase): Go to ArcGIS Online or Portal for ArcGIS Content > My Content page, click Add Item (see section 2.1).
- **Duplicate an existing feature layer without needing your own data:** This process will create a new empty layer of the same schema, in other words, containing the same attribute fields as the "mother" layer. This approach is useful if you can find a "mother" layer that matches your needs or if you want to use the "mother" feature layer again and again. You can go to ArcGIS Online or Portal for ArcGIS Content > My Content page, click Create > Feature Layer, and choose From Template, From Existing Layer, or From URL.
- **Create an empty feature layer and define your own fields interactively:** You will go to the ArcGIS for Developers site (**http://developers.arcgis.com**), log in with your ArcGIS Online account, click the plus button at the top, and select New Layer. The website will walk you through the steps to create a new feature layer. You will have the chance to specify the layer title, feature type (points, lines, or polygons), extent, attribute fields, and its default symbols.

Create an empty feature layer and define your own fields interactively using ArcGIS for Developers site.

Living Atlas of the World

Traditionally, you had to collect or prepare all or most of the data for your application and analysis. Today, you can find rich content from the ArcGIS Online catalog. You can use Living Atlas contents as your operational and, occasionally, basemap layers.

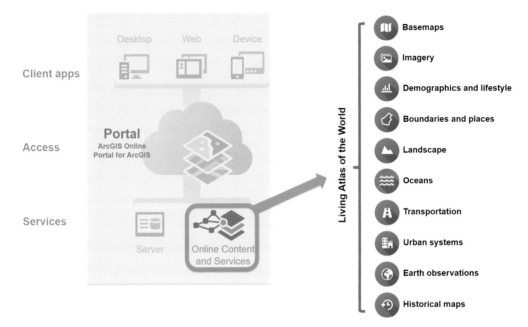

Living Atlas of the World is a curated subset of ArcGIS Online information items contributed to and maintained by Esri and the ArcGIS user community. The Living Atlas has thousands of layers covering many topics.

Living Atlas is a dynamic collection of thousands of maps, data, imagery, tools, and apps produced by ArcGIS users worldwide and by Esri and its partners. Living Atlas is the foremost collection of authoritative, ready-to-use global geographic information ever assembled. You can combine content from this repository with your own data to create powerful new maps and apps. You can use these maps and apps when you perform diverse analyses in ArcGIS Online, without having to collect the data yourself.

Living Atlas provides the following content categories:

- **Basemaps:** Reference maps for the world and the foundation for GIS apps.
- **Imagery:** Recent high-resolution imagery for the world, daily multispectral imagery, and near real-time imagery for major events, such as natural disasters.
- **Demographics and lifestyles:** Maps and data for the United States and more than 120 other countries reveal insights about populations and their behaviors.
- **Boundaries and places:** Boundaries help define where people live and work, and these layers span a variety of scales, from neighborhood to continental extents.

- **Landscape:** Data reflect both the natural environment and man-made influences to support land-use planning and management.
- **Transportation:** The collection of maps and layers reveals how people move between places.
- **Oceans:** Data on sea surface and floor temperature, distance to shore, ecological marine units, floor geomorphology, and various chemical concentration maps.
- **Urban systems:** More than half of the world's population now lives in cities, and these layers allow analysis of how population impacts the world.
- **Earth observations:** These observations depict our planet's extreme events and conditions, from severe weather to earthquakes and fires.
- **Historical maps:** These maps reflect the changing physical, political, and cultural aspects of our world over time.

The word "living" in the name Living Atlas indicates that content is continuously updated in minutes or hours (for example, live traffic and real-time earthquakes), days or weeks (for example, remote sensing imageries), and regularly as new content becomes available from the contributing community. For more information on the Living Atlas, go to **http://doc.arcgis.com/en/living-atlas/about**.

Layer configuration

Smart mapping

Your layers must be displayed in appropriate styles for you and your end users to discover the hidden patterns and deliver the intended messages. If your layers do not come with styles or you do not like their existing styles, you can change styles (although not all layers allow style changing) using ArcGIS smart mapping capabilities.

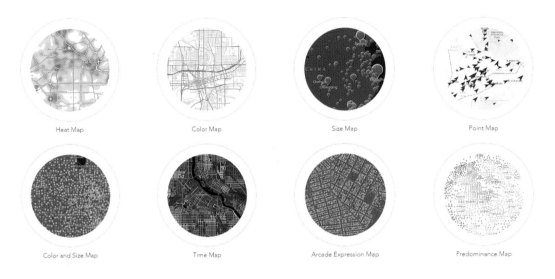

Heat Map   Color Map   Size Map   Point Map

Color and Size Map   Time Map   Arcade Expression Map   Predominance Map

Smart mapping aims to provide a strong, new "cartographic artificial intelligence" that supports map styles including a heat map, color map, size map, point map, color and size map, time map, predominance map, and more.

Smart mapping enables users to visually analyze, create, and share professional-quality maps easily and quickly with minimal mapping knowledge or software skills. Smart mapping uses a data-driven workflow to provide new and easy ways to symbolize your data and suggest the "smart" defaults. Smart mapping delivers continuous color ramps and proportional symbols, improved categorical mapping, heat maps, and new ways to use transparency effects to show additional details about your data via a streamlined and updated user interface. Unlike traditional software defaults that are the same every time, smart mapping analyzes your data quickly in many ways, suggesting the right defaults when you add layers and change symbolizing fields. The nature of your data, the map you want to create, and the story you want to tell all drive these smart choices.

Smart mapping doesn't just style your layer. It also performs exploratory data analysis to help you and your users discover the science of where. Smart mapping does not oversimplify the map-authoring experience or take control away from you. You can still specify parameters manually to extend default capabilities. For more information on smart mapping, see **http://www.esri.com/landing-pages/arcgis-online/smart-mapping**.

Pop-ups

Pop-up windows are considered windows to show geographic information and deliver geographic insight. They are a common tool that your end users rely on to interact with your operational layers. Today's users click or tap a location or feature on the map and expect to see a pop-up showing more information.

The default pop-up appearance for a layer is a plain list of attributes and values. You can configure the pop-ups to show custom-formatted text, attachments, images, and charts and links to external web pages. These pop-ups enhance the attributes associated with each feature layer in the map and present the information in more intuitive, interactive, and meaningful ways.

ArcGIS Arcade

How do you display data when there is no column in the schema containing that data? Traditionally you would have to alter the underlying data, such as adding a new column. But if you are not the owner or don't have the privilege, you can't alter the data. Arcade solves this common problem. Arcade is a portable, lightweight, and secure expression language written for use in the ArcGIS platform to style layers, label layers, and add values to layer pop-ups. With Arcade, you can build custom expressions based on existing fields and geometries without having to alter the data. Here are a couple simple examples:

- Label the percentage of the young and elder in the total population.
  ```
  Round(($feature.AGE_5DOWN + $feature.AGE_65UP)/ $feature.TOTAL_POP
  * 100 ) + "%"
  ```
- Calculate the weekday of the crime events in a layer.
  ```
  Weekday($feature.Reported_Date_Time)
  ```

You can almost think of Arcade as "Microsoft Excel for ArcGIS." In much the same way that Microsoft Excel lets you write formulas to work with spreadsheet data, Arcade lets you do the same with layer attributes and geometries. Like other expression languages, it can perform mathematical calculations, manipulate text, and evaluate logical statements. It also supports multi-statement expressions, variables, and flow control statements. Arcade was designed specifically for creating custom visualizations and labeling expressions in the ArcGIS Platform, including ArcGIS Online and ArcGIS Enterprise.

With Arcade, you can build custom expressions, use them to configure your layer style and pop-up without having to alter the underlying data.

While the syntax contains similarities to other scripting languages, Arcade is not designed for writing stand-alone apps. It is intended solely for evaluating embedded expressions such as those used in visualization, labeling, and alias contexts within applications of the ArcGIS platform. For more details, refer to **https://developers.arcgis.com/arcade/guide**.

Web app user experience design principles

User experience is an important factor when you configure your web layers, maps, and apps. A good Web GIS app should deliver informative content and enhance necessary functionality for a fast, easy, and fun user experience. Story Maps facilitate your creation of web apps with pleasant user experiences.

- **Fast:** "Don't make me wait," say today's users. Web GIS apps should use caching, database tuning, appropriate client/server task partitioning, and load balancing to achieve optimal performance, scalability, and availability. When you use ArcGIS Online, these are mostly taken care of by Esri for you automatically. But these aspects are important checkpoints when you use ArcGIS Enterprise to host your own layers and apps.
- **Easy:** Today's users also say, "Don't make me think about which button to click," and "If I don't know how to use your site, it's your problem, and I will leave the site quickly!" Web GIS apps should focus on a specific purpose. Do not overwhelm users with unnecessary buttons and data layers. Make the user interface intuitive. The interface should provide feedback, such as visual cues, that lead users through a well-defined workflow and assure them that they are on the right track.
- **Fun:** Integrate photos, charts, videos, and animation into your web apps. Used properly, these media enhance user engagement, convey your key information, and improve user satisfaction.

This tutorial

In this tutorial, you will learn how to create a Web GIS app that presents the spatial patterns of US population growth, explore the reasons behind the patterns, and share what you found with the public using Story Maps. In the process, you will learn the skills to create a feature layer using geocoding, style your layer using smart mapping, configure layer pop-ups using Arcade, and create a Story Map Journal web app.

    Data: For the operational layers, you are provided with a CSV file, **C:\EsriPress\GTKWebGIS\ Chapter2\Top_50_US_Cities.csv**, which contains the 2010 to 2016 population and housing information of the 50 most-populated cities in the US.

- There are no latitude and longitude fields in the CSV.
- Other operational layers are not provided. You will find them in the Living Atlas.

Requirements:
- Your map symbols should be easy to understand.
- If a city or region is clicked, your app should display associated details in intuitive ways.
- Your app should present the population change patterns and the reasons behind the patterns in story maps format.

System requirements:
- An ArcGIS Online or Portal for ArcGIS publisher account. You can use the trial account you created in chapter 1.
- A web browser.
- Microsoft Excel or a text editor.

2.1 Create a feature layer using geocoding

In this section, you will create a hosted feature layer, which will be your main operational layer.

1. In Microsoft Excel, navigate to **C:\EsriPress\GTKWebGIS\Chapter2\Top_50_US_Cities.csv**, and study the data fields.

Rank	City	State	Census2010	Estimate2010	Estimate2011	Estimate2012	Estimate2013	Estimate2014	Estimate2015	Estimate2016	OccHouseUnit	OwnerOcc	PopOwnOcc	SizeOwnOcc	RenterOcc	PopRentOcc	SizeRentOcc	Wikipedia_URL	Picture_URL
1	New York	New York	8175133	8192026	8284098	8361179	8422460	8471990	8516502	8537673	3,109,784	962,892	2,648,617	2.75	2,146,892	5,340,986	2.49	http://en.wikipedia.	http://upload.wikimedia.org/wik
2	Los Angeles	California	3792621	3796292	3825393	3858137	3890436	3920173	3949149	3976322	1,318,168	503,863	1,535,444	3.05	814,305	2,172,576	2.67	http://en.wikipedia.	http://upload.wikimedia.org/wik
3	Chicago	Illinois	2695598	2697736	2705404	2714120	2718887	2718530	2713596	2704958	1,045,560	469,562	1,264,635	2.69	575,998	1,370,717	2.38	http://en.wikipedia.	http://upload.wikimedia.org/wik
4	Houston	Texas	2099451	2105625	2132157	2166458	2204406	2243999	2284816	2303482	782,643	355,236	994,505	2.8	427,407	1,067,875	2.5	http://en.wikipedia.	http://upload.wikimedia.org/wik
5	Phoenix	Arizona	1445632	1450629	1469353	1499007	1525562	1554179	1582904	1615017	514,806	296,742	830,614	2.8	218,064	593,280	2.72	http://en.wikipedia.	http://upload.wikimedia.org/wik
6	Philadelphia	Pennsylvan	1526006	1528427	1539022	1550379	1555868	1560609	1564964	1567872	599,736	324,536	839,307	2.59	275,200	629,316	2.29	http://en.wikipedia.	http://upload.wikimedia.org/wik

Note: There are no latitude and longitude fields in the CSV file. The file contains the following fields:
- **Rank:** A city's 2016 rank by population.
- **City:** City name.
- **State:** Name of the state in which the city resides.
- **Census2010:** City population as of April 1, 2010.
- **Estimate2010, Estimate2011, Estimate2012, Estimate2013, Estimate2014, Estimate2015, Estimate2016:** Estimated population as of July 1 in 2010, 2011, 2012, 2013, 2014, 2015, and 2016.
- **OccHouseUnit:** Occupied house units in 2010.
- **OwnerOcc:** Owner-occupied house units in 2010.
- **PopOwnOcc:** Population in owner-occupied house units in 2010.

- **SizeOwnOcc:** Average house size of owner-occupied house units in 2010.
- **RenterOcc:** Renter-occupied house units in 2010.
- **PopRentOcc:** Population in renter-occupied house units in 2010.
- **SizeRentOcc:** Average house size of renter-occupied house units in 2010.
- **Wikipedia_URL:** URL to the city's Wikipedia page.
- **Picture_URL:** URL to the image of the city's seal or flag.

2. Open a web browser, navigate to ArcGIS Online (**http://www.arcgis.com**) or your Portal for ArcGIS, and sign in.

3. Click Content.

4. Click Add Item ╋, and then click From my computer.

5. In the Item from my computer window, perform the following tasks:

- For File, browse to **C:\EsriPress\GTKWebGIS\Chapter2\ Top_50_US_ Cities.csv**, and click it. If you have published a file of the same name to your content before, rename your Top_50_US_Cities.csv file to a unique name, and then select it.
- For Title, use the default, or specify a new one.
- For Tags, specify keywords, such as **US cities**, **Population change**, **GTKWebGIS**, and your organization name, as illustrated. Separate the keywords with commas.
- Make sure the check box next to Publish this file as a hosted layer remains selected.
- Leave the Address option selected.
- For Country, select US.
- Review the field types and location fields, and make corrections if needed.
- Click Add Item.

Item from my computer ❷   ✕

Add an item from your computer.

File:

| Choose File | Top_50_US_Cities.csv |

Title:

| Top_50_US_Cities |

Tags:

| US cities ✕ | Population change ✕ | GTKWebGIS ✕ | my organization name ✕ |
| Add tag(s) |

☑ Publish this file as a hosted layer. (Adds a hosted layer item with the same name.)

Locate features using:

○ Latitude/Longitude   ◉ Address   ○ None, add as table

Country:

| United States ▼ |

Review the field types and location fields. Click on a cell to change it.

Field Name	Field Type	Location Fields
City	String	City
State	String	State

Time Zone: | (UTC) Coordinated Universal Time ▼ | ⑦

[Add Item] [Cancel]

This CSV does not have latitude/longitude fields. ArcGIS Online and Portal for ArcGIS will geocode your data. Geocoding converts addresses and other identifiers (such as place-names and postal codes) into coordinates. If your data contains addresses for a single country, select its name from the Country list. If the addresses refer to multiple countries or to a country that is not on the Country list, select World. If ArcGIS Online didn't pick up the correct location fields, you can click the wrong fields to make corrections.

Note: Geocoding a CSV or a table is considered batch geocoding, which will cost you ArcGIS Online credits if you use ArcGIS Online's geocoding service. You can configure your ArcGIS Online and Portal for ArcGIS to use your own geocoding service.

6.   When prompted to review addresses, click Yes.

The feature layer will be added to the map viewer for you to review the geocoding results. You can click each of the matched addresses in the matched table at the bottom of the map and review its placement on the map. You can also click each of the unmatched addresses in the unmatched table, fix it by choosing a suggested address, or Edit the address information and choose a suggested address, or directly add a point to the map.

In the tutorial data, all cities are matched to their correct locations.

7.   Click Done Reviewing.

You have created a hosted feature layer using geocoding.

8.   Click the Save button to save your web map.

9.   In the Save Map window, enter the appropriate title and tags, and then click Save Map.

You have created a hosted feature layer using geocoding. You can go to the layer details page to share the layer with everyone or other audience you select. In this tutorial, you will share the layer together with your web map later.

2.2 Configure layer style

In this section, you will explore smart mapping and configure your layer to display the population change rates in the 50 cities in your layer.

1. With the map viewer open, in the Contents pane, point to Top 50 US Cities Population, and then click the Change Style button 🖼 .

2. In the Choose an attribute to show text box, choose Estimate2016 as the field to show in the Change Style pane.

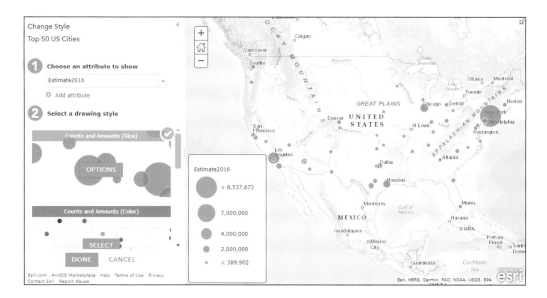

Note that smart mapping automatically selects Counts and Amounts (Size) as the default style for this numeric field. This style uses an orderable sequence of different sizes to represent your numerical data or ranked categories. You can symbolize points, lines, and areas using this approach. The proportional symbols make the map intuitive. On this map, can you easily tell the top three most populated cities in the US?

Next, you will style the layer based on the ratio of two fields. You can do so using Arcade or selecting two fields. Here, you will select two fields.

3. Click Options in the Counts and Amounts (Size) style to further configure the layer style.

4. Set Divided By to Census2010.

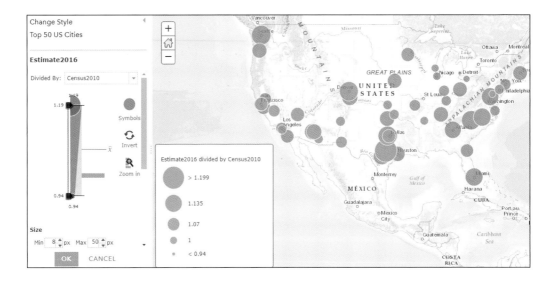

The map now displays the ratio of population 2016 to 2010, which is a good indicator of population growth over the years. Bigger circles indicate bigger increases, and smaller circles indicate smaller increases and even decreases. A value of 1.0 means there was no population change; numbers greater than 1.0 indicate population growth; and numbers less than 1.0 indicate population decline. This ratio is a good approach to reveal how things change over time without having to create multiple maps. This approach works for any numerical data from two time periods.

You may also change the Min and Max symbol sizes and adjust the two handles on the histogram slider to exaggerate the cities with the most and the least population growth. Smart mapping easily emphasizes certain ranges of the data and uncovers subtle details.

Smart mapping also uses continuous sizes and colors automatically, but it does not remove any of the traditional methods to break features into a set of classes and show them with a limited number of sizes and colors. Next, you will manually break the cities into classes and assign symbols for these classes.

5. Scroll down in the Change Style pane and select the box next to Classify Data. Select manual breaks with 4 classes and Round classes to 0.01 (in other words, round the class break values to 2 decimal points).

☑ Classify Data

    Using | Manual Breaks | ▼

    With | 4 ▲▼ | classes

    Round classes: | 0.01 | ▼

6.   Starting from the bottom, set the class break values to **1**, **1.1**, and **1.15**. You can set the values by clicking the existing break values and typing in the new values as illustrated or by dragging the handles on the histogram slider in Change Style.

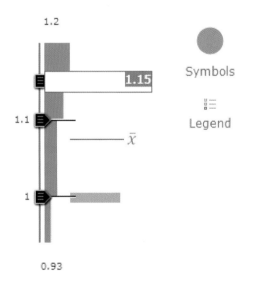

Note the Symbols and Legend icons in the figure. Clicking the Symbols icon allows you to set a symbol for all the classes. Clicking the Legend icon allows you to set an individual symbol and label for each class.

All the classes are using the same symbol with gradual sizes. Next, you will set each class to use a different symbol.

7.   Click Legend.

This step allows you to edit the symbol for each individual class.

8.   Click the biggest circle, in other words, the one next to > 1.15 to 1.2.

A window appears showing the available point symbols.

9. After clicking the list to choose Arrows, click the solid red up arrow, set its size to **36**, and click OK.

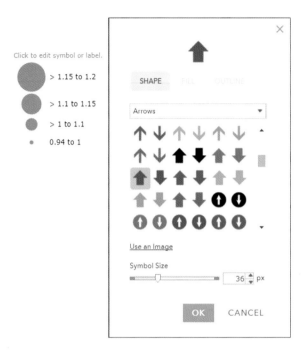

The map viewer provides a collection of symbols for business, damage, disasters, infrastructure, recreation, people places, points of interest, health, transportation, and other categories, as shown in the list of options. You can browse through these categories and get a sense of the variety of available symbols. To browse, you can click the symbol list to access the options.

10. Similarly, make the following changes:

- For the > 1.1 to 1.15 class, change the symbol to a brown upper arrow in the Arrows group, set its size to 24, and click OK.
- For the > 1.0 to 1.1 class, change the symbol to a purple upper arrow in the Arrows group, set its size to 16, and click OK.

In addition to the collections of symbols, the map viewer allows you to use your own images as symbols. You will do so in the next steps.

11. Click the symbol for the 0.94 to 1 class, click Use an Image, specify **https:// i.imgur.com/OS8tZpq.gif** (or the short equivalent **http://bit.ly/2F0fQVq**) as the URL, click the Plus icon, set its size to **16**, and then click OK.

The last class lost population from 2010 to 2016, so the class was symbolized using a solid down arrow.

12. Click OK, and then click Done to exit the Change Style pane.

13. In the Contents pane, point to the Top 50 US Cities layer, click the More Options button ···, and click Save Layer.

This step saves the style to the layer item itself. When you add this layer to any web maps, the layer will carry the style you defined in this section as the default style.

Next, you will change the basemap to a neutral background.

14.   Click the Basemap button ⊞ on the map viewer toolbar, and click Light Gray Canvas.

This basemap provides a neutral background with minimal colors, labels, and features. This background allows your operational data layer to stand out clearly.

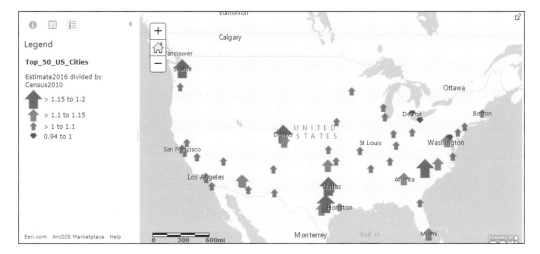

15.   Save your web map.

You have configured the style of your operational layer. With this map, you can quickly and easily see which cities had population increases or decreases.

2.3 Configure a layer pop-up using ArcGIS Arcade

In this section, you will decide what information to show and how you will show the information in the cities' pop-ups.

1.   With your web map open in the map viewer, click one of the 50 cities to see the default pop-up.

If a pop-up window does not appear, it may be disabled for this layer. You can enable the pop-up by pointing to the layer, clicking the More Options button and clicking Enable Pop-up.

This default pop-up is useful, but you can enhance the window to communicate information in more intuitive and engaging ways.

2. In the Contents pane, point to the Top 50 US Cities layer, click the More Options button ••• , and click Configure Pop-up.

The Configure Pop-up pane appears, in which you can configure the pop-up's title, contents, and media sections. Next, you will configure the title, which can be static text, a set of attribute field values, or a combination of the two.

3. Click the Plus button ⊞ under Pop-up Title to select the City field. Type a comma and a space, and click the Plus button again to select the State field.

Configure Pop-up

Top_50_US_Cities

☑ Show Pop-ups

Pop-up Title

{City}, {State}   ⊞

Rank {Rank}
City {City}
State {State}
Census2010 {Census2010}

Next, you will configure the pop-up contents, which can include a list of attribute fields or a custom description based on attribute values. You will first create a couple attribute expressions using Arcade.

4. Under Attribute Expressions, click the Add button.

Configure Attributes

Attribute Expressions

Adding expressions allows you to create new information from existing fields for use in pop-ups.

ADD

The expression editor window appears. Next, you will build the expression `Round($feature .Estimate2016 / $feature.Census2010, 2)`, which calculates the ratio of population 2016 and 2010 for each city, and rounds the result to two decimals.

5. In the expression editor, do the following:

- Click the Edit link by the title Custom and change the title to **Population Ratio 2016/2010**.
- Click in the Expression text area.
- Click Functions, search and click Round.
- In the Expression area, remove fieldOrValue.
- Click Globals, find and click $feature.Estimate2016, press the / key, and find and click $feature.Census2010.
- Click numPlaces and change it to **2**.
- Click Test to verify if the results look correct. If not, correct the expression based on the error message.
- Click OK.

Population Ratio (2016/2010) ✎ Edit

| Expression | | Test | Globals | Functions | ⓘ |

```
1  // Write a script that returns a value that will be used to draw features.
2  // For example, find the percentage of males:
3  // Round(($feature.MalePop / $feature.TotalPop) * 100, 2)
4  Round($feature.Estimate2016 / $feature.Census2010, 2)
```

$feature.Estimate2014

Field: Estimate2015
$feature.Estimate2015

Field: Estimate2016
$feature.Estimate2016

Field: FID
$feature.FID

Field: OccHouseUnit
$feature.OccHouseUnit

Field: OwnerOcc
$feature.OwnerOcc

Field: Picture_URL
$feature.Picture_URL

Field: PopOwnOcc
$feature.PopOwnOcc

Field: PopRentOcc

Results   Messages

Result	Value
Result	1.04
Type	Number

OK   Cancel

Next, you will build an expression like Round($feature.PopOwnOcc / $feature.Census2010, 2) * 100, to calculate the percentage of population-owned housing in 2010 for each city.

6.   Under Attribute Expressions, click the Add button again. In the expression editor, perform the following actions:

   • Change the title to **% Population Own Houses**.
   • Set the expression to **Round($feature.PopOwnOcc / $feature .Census2010, 2) * 100**.
   • Click Test to make sure the results look correct.
   • Click OK.

7.   In the Display menu under Pop-up Contents, make sure A list of field attributes is selected, and then click the Configure Attributes link.

Pop-up Contents

Display: A list of field attributes   ▾

These field attributes will display:

Rank {Rank}
City {City}
State {State}
Census2010 {Census2010}   ⇧
                                            ⇩

Configure Attributes

   • In the Configure Attributes dialog box, make the following choices:
   • Select {Rank} and set its alias as **Rank 2016**.
   • Select the expression field Population Ratio (2016/2010).
   • Select the expression field % Population Own Houses, and set its format to 0 decimal places.
   • Clear the check boxes for the rest of the fields.

Some of the fields already appear in the title, and some others will be displayed in charts and image links. Thus, you unselect these fields.

   • Hiding them from the attribute list will make your layer pop-up simple and clear.
   • Click OK to close the Configure Attributes window.

8.   At the bottom of the Configure Pop-up pane, click OK to apply your pop-up configuration.

9.   Click any city on the map to review the new pop-up.

You will see that the pop-up window is simpler and easier to read than the default pop-up. Also note the pop-up includes the custom expression you built with Arcade.

10.   On the map viewer menu bar, click Save to save your changes.

2.4 Add images and charts to pop-up windows

Media, such as images and charts, can more effectively engage users and improve their understanding of your data.

The tutorial CSV contains two URL fields, Wikipedia_URL and Picture_URL. You will use the first URL to add a picture to the city layer's pop-up and the second URL to link the picture with the city's Wikipedia page so that users can gather supplementary information about the city's population changes. Charts require numeric attribute fields. The US cities layer contains several of these fields. You will display them in appropriate charts to exemplify the cities' population trends.

1.   In the Contents pane, point to the Top 50 US Cities layer, click the More Options button, and click Configure Pop-up.

2.   In Pop-up Media, click Add, and then click Image.

Pop-up Media

Display images and charts in the pop-up:

ADD ▾

Image	harts.
Pie Chart	ld one.
Bar Chart	to order.
Column Chart	
Line Chart	

In the Configure Image window, make the following changes:

• Enter **About the city** as the title (leave the Caption window blank).
• In the URL box, click the Plus button, and click the Picture_URL field.
• In the Link (optional) box, click the Plus button and click the Wikipedia_ URL field.

3.   Click OK to close the Configure Image window.

The image title, caption, image URL, and link URL can all take the form of static text, attribute field values, or a combination of the two. If you do not have image and link URL fields when you do the assignment, specify a static URL instead. For example, you can use **http://www.census.gov/ history/img/Census_Logo.jpg** as the image URL, and **http://www.census.gov** as the Link URL. This way, the pop-up windows for all cities display the same image and link to the same web page. You can add additional images to your pop-up window simply by repeating steps 2 and 3.

4. In the Configure Pop-up pane, click the OK button to apply your pop-up configuration.

5. Click a city on the map to observe the new pop-up.

The pop-up window displays the city's seal or flag. If you click on the image, the city's Wikipedia page will appear.

6.   In the Contents pane, point to the Top 50 US Cities layer, click the More Options button, and click Configure Pop-up again. In the Pop-up Media section, click Add, and then click Line Chart.

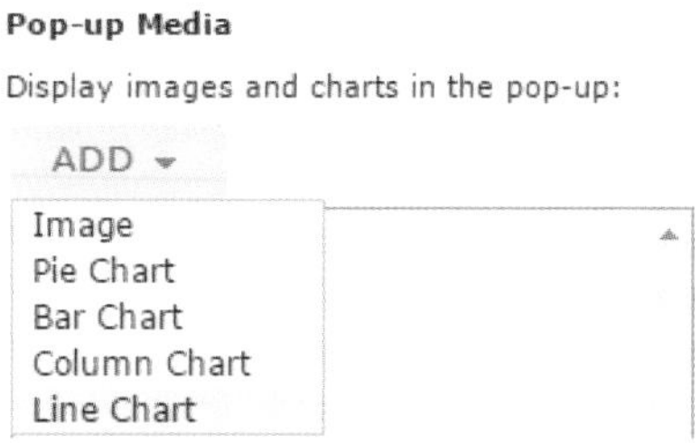

Pop-up Media

Display images and charts in the pop-up:

ADD ▾

| Image |
| Pie Chart |
| Bar Chart |
| Column Chart |
| Line Chart |

7.   In the Configure Line Chart window, make the following settings:

- For Title, specify **Population Change (2010 - 2016)**.
- For Chart Fields, check the 2010, 2011, 2012, 2013, 2014, 2015, and 2016 population fields.

8.   Click OK to close the Configure Line Chart window

Configure Line Chart   ×

Specify the title, caption and fields to chart.

Title:

| Population change (2010 - 2016) | ⊞ |

Caption

| | ⊞ |

Chart Fields

☐ Field Alias	Field Name
☐ Rank 2016	{Rank}
☑ Census2010	{Census2010}
☑ Estimate2010	{Estimate2010}
☑ Estimate2011	{Estimate2011}
☑ Estimate2012	{Estimate2012}

Normalize by: None ▾

OK   CANCEL

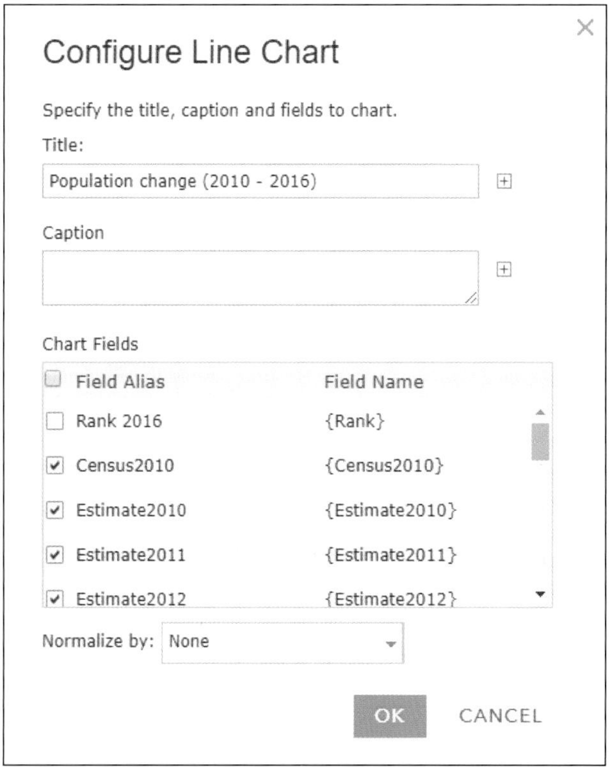

9.   Click OK to apply your configuration.

10. Click a city, for example, Detroit, Michigan, to see the new pop-up (use the
     Search box to find Detroit and other cities if necessary). In the pop-up, to the
     right of the city seal or flag image, click the right arrow ▶ to see the chart you
     configured.

You will notice that the population of Detroit has been decreasing.

11. In the Contents pane, point to the Top 50 US Cities layer, click the More
     Options button •••, and click Save Layer.

This step saves the pop-up configuration to the layer item itself. When you add this layer to
any web maps, the layer will carry the pop-up you defined in this section as the default.

12. Save your web map.

The new pop-up you configured does more than display raw attribute values. The pop-up
leverages the URLs stored in the attributes to display and link a picture to a website that provides
additional details. The pop-up also charts the numeric fields to provide a visual interpretation of
population changes.

2.5 Use layers from the ArcGIS Living Atlas of the World

You don't always have to create your own layers. You can use layers from the Living Atlas as your operational layers. In this section, you will use a layer from the Living Atlas to further display the US population change patterns.

1. On the map viewer menu bar, click the Add button ⊞ , and from the list, click Browse Living Atlas Layers.

In the Living Atlas pane, you will explore the extensive categories and subcategories of layers available.

2. In the Living Atlas pane, click the Filter button to see the list of categories.

3.   In the Filter pane, browse the Categories, Regions, and Item Type groups. Notice the options available, and click the X button to close the Filter pane.

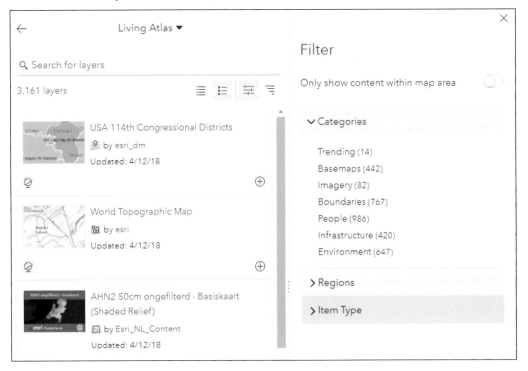

4.   In the search box, type **population growth**, and press Enter.

5.   In the result list, click the 2017-2022 USA Population Growth layer to see the Item Details pop-up. Click Add to map, and then close the Item Details pop-up.

If you added the wrong layer, you can remove that layer by clicking the Remove button.

6.   In the Contents pane, point to the 2017-2022 USA Population Growth Rate layer you just added, click the More Options button ••• , and click Show Item Details.

This action will bring up the item details about the layer. This map layer shows the estimated annual growth rate of the population in the US from 2017 to 2022 in a multiscale map by state, county, ZIP Code, tract, and block group.

7. Go back to the map viewer by clicking the corresponding browser tab. Zoom in and out of the map to examine the population change patterns. Click a state, county, zip, tract, or block group to see the information in the pre-configured pop-ups.

You should see that the population change trends of the 50 major cities generally agree with the trends of their counties, zips, tracts, and block groups.

Note: You may see multiple features at the location you clicked. In such cases, the pop-up will show one feature at a time. You can click the arrow in the pop-up header to navigate through the features. You may also see pop-ups of multiple layers. You can disable the pop-up for a layer in the layer's more options context menu.

8. Zoom the map to show all the 50 cities and click Save to save your web map.

The map extent saved is the default extent of your web map.

9. Click the Share button ⊕⊟.

10. In the Share window, share your web map with everyone (public).

11.   When prompted to update the sharing level of your layer in the web map to be the same as the web map, click the Update Sharing button.

Update Sharing

×

These layers in the web map may not be visible to others because they are not shared in the same way as the web map.

Layer	Owner
Top_50_US_Cities	GTKWebGIS

Click Update Sharing to adjust the settings of the layers you can update so they can be viewed in the web map.

UPDATE SHARING   CANCEL

You have configured your layer's style and pop-up, and used the layer in a web map. Next, you will use this web map in your web app.

2.6 Create an Esri Story Map Journal web app

In this section, you will create an app to showcase the US population change pattern and the reasons behind it. To keep the tutorial short, a side-by-side comparison web app and a swipe comparison web app have been created for you.

In the process, you will embed two pre-created story maps into the journal.

1.   In a web browser, go to the Story Maps website (**https://storymaps.arcgis .com**), sign in with your ArcGIS Online account, and in the top menu, click Apps.

2.   Find Story Map Journal and click Build.

A Rich Multimedia Narrative

Story Map Journalˢᴹ

Create an in-depth narrative organized into sections presented in a scrolling side panel. As users scroll through the sections in your Map Journal they see the content associated with each section, such as a map, 3D scene, image, video, etc.

BUILD   OVERVIEW   GALLERY   TUTORIAL   FAQS

3.   In the Welcome window, select Side Panel, and click Start.

You will be directed to the map journal builder.

4.   Enter your map journal title, type **Story of US Population Change**, and click the Continue button.

The **Add Home Section** window appears. A Story Map Journal is made up of sections. The home section will act as the cover page of your story. Each section has a main stage and a side panel.

5.   For Step 1: main stage content, set these parameters:

- For Content, select Web page.
- For Webpage URL, type **https://www.census.gov/popclock**, and click Configure.
- Leave Fill as the position choice.
- For Load page over a secure connection (HTTPS), unselect the check box if your current page uses HTTP; otherwise, select the check box.
- Click Next.

The Census webpage supports both HTTP and HTTPS. Remember that web browsers don't allow mixed HTTP and HTTPS contents. If your current page URL uses HTTPS, the web page you want to use here must also support HTTPS and you must use the HTTPS URL. If the web page you want to use doesn't support HTTPS, you need to reload your current page using HTTP. Refer to chapter 1, "Questions and answers," regarding HTTP and HTTPS settings.

The main stage is configured. The builder now leads you to the side panel configuration.

6.   In the side panel text area, type **Population in the world and in the US are constantly changing**.

7.   Click Add.

The home section displays.

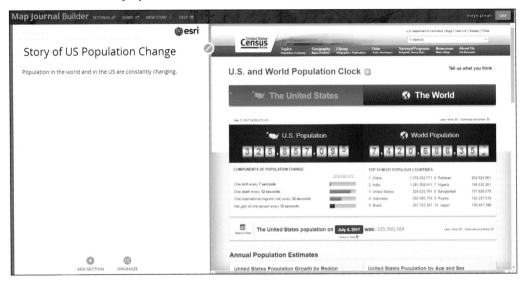

8.   Click Save (the button at the upper-right corner) to save your app.

Save your app after adding each section to avoid losing your work accidently.

9.   Click Add Section. In the Add Section window, select the following settings:

 • For Section title, type **Not all cities change in the same way**.
 • For Content, leave Map selected.
 • For Map, choose Select a map.

ADD SECTION

STEP 1: Main Stage Content

Not all cities change in the same way

CONTENT: ● Map   ○ Image   ○ Video   ○ Web page

Map   Select or create a map

   Select a map
   Create a map

NEXT   CANCEL

10. In the Select a map window, select the US Population Change web map you created in section 2.5.

11. If you don't have that web map, search in ArcGIS Online for "chapter2 population change owner:GTKWebGIS," and then select the found web map.

Select a map		x
ArcGIS Online	▾	ɔopulation change owner:GTKWebGIS ⌐

US Population C...
GTKWebGIS

1 - 1 of 1 results

You can configure how you want the web map to be shown, including the location, content, pop-up, and extras. Here you will leave them as the default.

12. Click Next.

13. For Step 2: Side Panel Content, type **Can you tell several cities with the most population increase and decrease?** in the text area.

14. Click Add.

Section 2 is added.

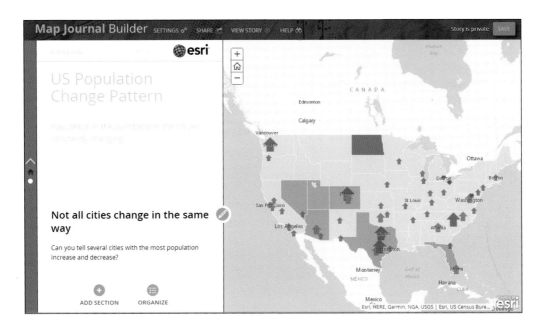

15. Click Add Section. In the Add Section window, set the following parameters:

- For Section title, type **More discoveries**;
- For Content, select Web page;
- For Webpage URL, type **https://arcg.is/2xWlvaj**, click Configure, leave Fill as the position choice;
- For Load page over a secure connection (HTTPS), clear the check box if your current page uses HTTP, otherwise, select the check box;
- Click Next.

16. For Step 2: Side Panel Content, in the text area, type the following text:

Cities with more population increase tend to have

- **more people own housing than renting on the predominant map**
- **owners occupy relative larger housing size on the house size map**
- **lower unemployment rates**

17. Click Apply.

18. Click Add.

Next, you will add a new section to summarize your story.

19. Click Add Section. In the Add Section window, do following:

    - For Section title, type **Why**;
    - For Content, select Web page;
    - For Webpage URL, type **https://bit.ly/2GZhdov**, click Configure, leave Fill as the position choice;
    - For Load page over a secure connection (HTTPS), clear the check box if your current page uses HTTP, otherwise, select the check box;
    - Click Next.

20. For Step 2: Side Panel Content, click the Insert an image, video, or web page button 📷◼ , and click Upload.

21. Browse to **C:\EsriPress\GTKWebGIS\Chapter2\images\reasons-why -people-move.jpg**, leave the Image caption as is, and click Apply.

22. Click Add.

    The section is added.

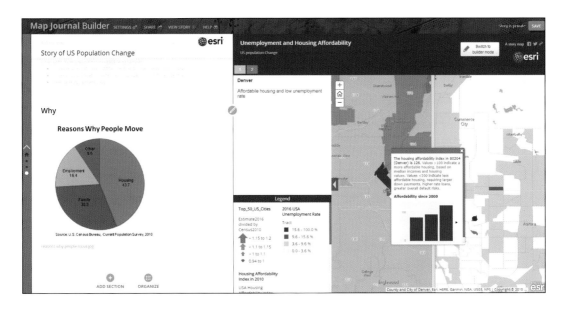

The pie chart image shows that people move because of housing, family, and employment. The app on the right is an Esri® Story Map Swipe℠ app. The app lets users compare the housing affordability layer and the unemployment rate layer by sliding the swipe bar back and forth. Refer to the "Questions and answers" section for how to create a swipe web app.

The swipe web app has bookmarks defined. The first bookmark, which is the default map extent of the swipe web app, is Denver.

23. Click areas in Denver to see the housing affordability index in the pop-up, then move the vertical bar in the swipe web app to see the unemployment info in the pop-up.

The Denver area has relatively high housing affordability values and low unemployment rates, which are the main reasons for its population increase.

24. Click bookmark 2 in the swipe web app.

The Detroit and Cleveland areas have high unemployment rates, which is the main reason for their population decreases.

25. Click Save to save your app.

26. Click Share in the top menu bar.

27. In the share window, click Public.

If you see a warning message that reads, "There are issues in your story content that will be noticeable to your readers," the warning message often indicates that this story map contains premium content. You can click the Ignore button, which implies that users will need to sign in to ArcGIS Online and consume their credits to see your app. Or you can click the Make Public button (you may need to scroll down the window to see the button) to allow users to view the premium content using your credits.

28. Click View Story.

Your app will open in a new window or new tab.

29. Navigate through your map journal from the cover to the end. At each section, interact with the contents on the main stage.

At each step, you can imagine how you would tell your story.

30. Share the URL of your app with your audience.

The URL of the current web page is the URL of your app. You can also go back to the configuration mode, click the Share button, and share your app via Facebook and Twitter. You can even embed your story map app into your blogs or other web pages.

In this tutorial, you created a feature layer using geocoding; configured its style using smart mapping; configured its pop-up using Arcade; added images, links, and charts to the pop-up; and experienced the flexibility of smart mapping and the rich contents of the Living Atlas. You then created a Story Map Journal web app, embedded a web page, an image, a web map, a side-by-side-comparison web app, and a swipe web app. These materials presented the population growth pattern and the possible reasons behind the pattern clearly.

QUESTIONS AND ANSWERS

1. After I created my web app using my web map, I updated the layer styles and pop-ups of my web map. Will the changes be reflected automatically in my web app?

   Answer: Yes.
   A web app maintains a link with its web map and depends on the web map as its source. Changes made to web map content, such as adding or removing layers, changing symbology, or configuring pop-up windows, will reflect automatically in the app.

2. CSV files can only hold point features. How can I create a line or polygon feature layer?

   Answer: You can use shapefiles, file geodatabases, or enterprise geodatabases.
   - Just zip the files in your shapefile or file geodatabase into one zip file (a shapefile is an Esri vector data storage format that contains .shp, .shx, .dbf, and .prj files, and a file geodatbase is a folder containing many files), and go to Content > My Content > Add Item > From my computer, browse to the zip file, and follow the instructions.
   - If you have layers in your enterprise geodatabase, you can use ArcGIS Pro or ArcMap to publish them into ArcGIS Online or ArcGIS Enterprise.

3. Can I insert video or audio clips into pop-ups? How?

   Answer: Yes, use the custom attribute display and HTML source as illustrated.

4. Can I write HTML code directly in story maps?

   Answer: Yes. While configuring Map Journal side panel contents, you can switch the text edit area to the HTML source code editor by clicking the Source icon ⬚ , and write HTML code there. Most other story maps support HTML code in similar ways. This capability gives developers great freedom in customizing the contents. For example, you can add audio clips to your story as illustrated. Refer to **http:// arcg.is/2wX5nl5**.

5. Can I add a 3D scene to my story map?

   Answer: Yes, refer to **http://arcg.is/2wjRMqv**.

6. What is a predominant map, and how can I create a predominant map in ArcGIS Online?

   Answer: A predominant map compares attributes that share a common subject and unit of measurement to see which has the highest value. For example, an election map can use it to display which one of the two or more candidates won the most votes, and by how much. When styling a layer in the map viewer, choose multiple numeric columns, and then you will see Predominant Category and Size styles. Choose one and follow the instructions to create a predominant map.

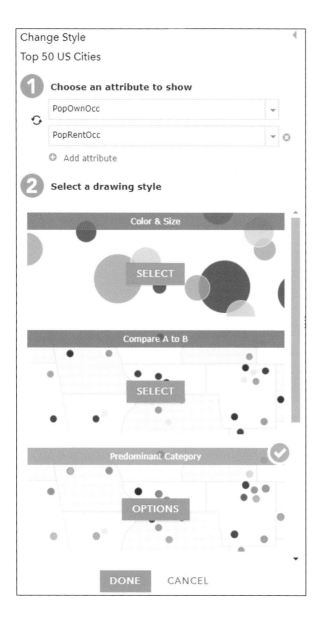

7. How do I create a Story Map Swipe web app?

   Answer: First, create a web map that includes the two layers you want to compare. Second, go to **http://storymaps.arcgis.com/en/app-list**, find Story Map Swipe, click Build and follow the instructions.

ASSIGNMENT

Assignment 2: Create a hosted feature layer, and use it in a story map to tell a story about the following topics:

- Your personal life or passion
- The spread of a virus
- A historical war or other event
- Spatial patterns of various social or natural factors
- The real estate for sale in your city, along with pictures and a chart of price histories
- The recent crime sites in your city
- The recent earthquakes in the world
- Other topics of your choosing

Data:
- Use data from the Living Atlas, ArcGIS Open Data, or other sources. You can certainly make up your own data too.

Requirements:
- Your layers should have appropriate styles.
- Configure appropriate layer pop-ups. Include at least one chart or image with a link in the pop-up. (If there are no image URLs in your data, find an image URL.)
- Use Arcade when styling your layer or configuring your layer pop-up.
- Use map journal, map cascade, map series, or map shortlist story map format.
- Include five types of resources: text, video, picture, map, and web page.

What to submit:
- The URL to your web app

Resources

"Add <audio> to your Map Journal," Owen Evans, https://developerscorner.storymaps.arcgis.com/using
-the-html-audio-tag-in-your-story-map-f818d9316252 (or http://arcg.is/2wX5nI5).

"Add Layers," https://doc.arcgis.com/en/arcgis-online/create-maps/add-layers.htm.

"ArcGIS Arcade," https://developers.arcgis.com/arcade.

"ArcGIS Online: Data Exploration with Smart Mapping," Jennifer Bell and Lisa Berry, https://www.
youtube.com/watch?v=zlKcR1HlrP0 (or http://bit.ly/2EGPyWZ).

"ArcGIS Online Steps for Success—A Best Practices Approach," Bern Szukalski and Jeff Archer, https://
www.youtube.com/watch?v=qUIIM9HrfRU (or http://bit.ly/2AYLVJH).

"Change style," https://doc.arcgis.com/en/arcgis-online/create-maps/change-style.htm.

"Chapter 2, Mapping Is for Everyone," *The ArcGIS Book*, Christian Harder and Clint Brown, http://learn
.arcgis.com/en/arcgis-book/chapter2.

"Chapter 3, Tell Your Story Using a Map," *The ArcGIS Book*, Christian Harder and Clint Brown, http://learn
.arcgis.com/en/arcgis-book/chapter3.

"Configure pop-ups," https://doc.arcgis.com/en/arcgis-online/create-maps/configure-pop-ups.htm.

"Create apps from maps," https://doc.arcgis.com/en/arcgis-online/create-maps/create-map-apps.htm

"Creating Story Maps with ArcGIS," https://www.esri.com/training/catalog/5887d359378ef44b34da22b2/
creating-story-maps-with-arcgis (or http://arcg.is/2nn7xK7).

"Esri UC 2017: ArcGIS Online Map Viewer—Do You Know?" Jennifer Bell, https://www.youtube.com/
watch?v=iPN2imGyXlI (or http://bit.ly/2mwPTns).

"Esri UC 2017: ArcGIS Online Map Viewer—Do You Know?" Bern Szukalski, https://www.youtube.
com/watch?v=9RelgSyqUHo&t=3s (or http://bit.ly/2mFp5mf).

"Fight Child Poverty with Demographic Analysis," http://learn.arcgis.com/en/projects/fight-child-
poverty -with-demographic-analysis (or http://arcg.is/2msQKpq).

"Get Started with ArcGIS Online," http://learn.arcgis.com/en/projects/get-started-with-arcgis-online.

"How to Smart Map in 3 Easy Steps," Lisa Berry, https://blogs.esri.com/esri/arcgis/2016/12/01/
how-to-smart-map-in-3-easy-steps.

"Inform and Engage Your Audience with Esri Story Maps," Bernie Szukalski and Rupert Essinger, https://
www.esri.com/training/catalog/57d876188b3e1ff2376c1539/inform-and-engage-your-audience-with
-esri-story-maps (or http://arcg.is/2cKAqME).

"Introducing Arcade," Paul Barker, https://blogs.esri.com/esri/arcgis/2016/12/19/introducing-arcade.

"Living Atlas Community Webinar—February 2017," Jim Herries, Lisa Berry, Jennifer Bell, and Tamara
Grant, https://www.youtube.com/watch?v=lfo10lCFy9c (or http://bit.ly/2Dw5Gev).

"Living Atlas of the World," http://doc.arcgis.com/en/living-atlas/about.

"Short videos, ArcGIS Online Help," http://doc.arcgis.com/en/arcgis-online.

"Story Map Journal source code," https://github.com/Esri/storymap-journal.

"Story Maps Gallery," http://storymaps.arcgis.com/en/gallery.

"Story Map Swipe and Spyglass source code," https://github.com/Esri/storymap-swipe.

"Using Living Atlas subscriber content in public maps and apps," Bern Szukalski, https://blogs.esri.
com/esri/arcgis/2017/03/14/living-atlas-subscriber-content.

"What's New in Arcade," Paul Barker, https://blogs.esri.com/esri/arcgis/2017/06/28/whats-new-in
-arcade-june-2017.

CHAPTER 3

Web AppBuilder for ArcGIS

Have you run into situations where you need more functions than any individual configurable app can provide? Do you wish you could remix the functions of multiple apps? For many users and in many situations, the answers to these two questions are yes. Web AppBuilder for ArcGIS is intended for such needs. It provides more functionality than any other ArcGIS configurable web app or template and is more flexible and configurable. Web AppBuilder comes with more than 30 premade widgets covering functions including mapping, table view, querying, charting, reporting, routing, geoprocessing, and more. The user community also can create more custom widgets. Web AppBuilder allows you to create web apps by selecting, mixing, and configuring widgets interactively—all without programming. Web AppBuilder also provides many themes (in other words, styles and layouts) for you to create easy-to-use, friendly, and responsive user interfaces that work for desktop, tablet, and mobile devices.

Learning objectives

- *Know why and when you need Web AppBuilder for ArcGIS.*
- *Understand the types of widgets and themes of Web AppBuilder.*
- *Learn the workflow to create web apps using Web AppBuilder.*
- *Configure and use charting, filtering, and other various widgets.*

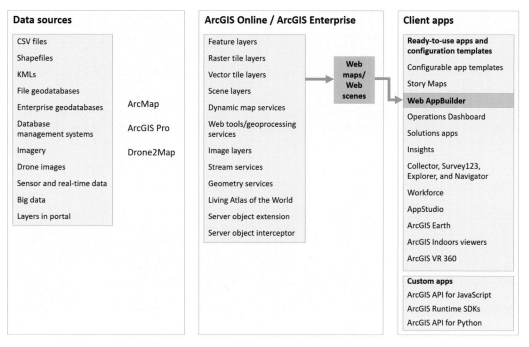

ArcGIS offers many ways to build web applications. The green lines in the figure highlight the technology presented in this chapter.

Web AppBuilder for ArcGIS and its basics

Web AppBuilder for ArcGIS is an intuitive what-you-see-is-what-you-get (WYSIWYG) application that allows you to build 2D and 3D web apps without writing a single line of code. The application includes powerful tools to configure fully featured HTML apps. As you add your map and tools, you can see them in the app and use them right away. Web AppBuilder is based on ArcGIS Online, ArcGIS Enterprise, HTML5, and ArcGIS® API for JavaScript™ technologies and has the following key features:

- Creates pure HTML and JavaScript apps that do not require any plug-ins.
- Uses responsive web design technologies to create web apps that work well on desktops, tablets, and smartphones.
- Comes with numerous out-of-the-box widgets (you can use the widgets immediately) that can be flexibly remixed and configured.
- Has a collection of configurable themes so you can customize the look and feel of your apps.
- Provides an extensible framework for developers to create custom widgets, themes, and applications.

Web AppBuilder allows users to choose from the available user interface themes, web maps, and widgets in a "what-you-see-is-what-you-get" designer experience. You can immediately see how your app will look as you change the configuration.

Editions of Web AppBuilder for ArcGIS

The Web AppBuilder product family has three editions:

- Embedded in ArcGIS Online
- Embedded in Portal for ArcGIS
- Developer Edition

Although the first two editions are embedded, generally, all three editions have similar functionality. For example, they will have the same designer user experience to create web apps, and similar widgets and themes. However, the detailed functionality of the three editions is not equivalent. Typically, new enhancements to Web AppBuilder are added first to the ArcGIS Online edition, then to the Developer Edition, and finally to the Portal for ArcGIS edition. Therefore, the editions may differ in themes, widgets, and other aspects during certain periods. Another difference among these editions lies in their support for custom widgets and themes.

- The ArcGIS Online edition doesn't allow users to use custom widgets.
- The Portal for ArcGIS edition allows users to use custom widgets.
- The Developer Edition allows users to create and use custom widgets.

The chapter tutorial is based on the edition embedded in ArcGIS Online, but the skills you will learn apply to all three editions.

Access to Web AppBuilder

For the embedded editions, you can access Web AppBuilder from the ArcGIS Online and Portal for ArcGIS Map Viewer, Gallery, or Content.

1. If you opt to use Map Viewer, you would click Share, click Create a Web App, and then click the Web AppBuilder tab (see section 3.2).
2. For the second option, you would choose Content > My Content > Create > App, and then click Using the Web AppBuilder.

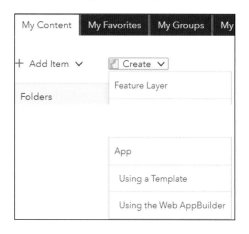

3. If you start from the Gallery, you would go to the Esri Featured Content > Apps section, and choose App Builders.

Using Web AppBuilder to create web apps

You can create a web app using Web AppBuilder for ArcGIS by following these steps:
- **Pick style:** Configure the look and feel of the app by picking a theme. A theme includes a collection of panels, styles, layouts, and pre-configured widgets.
- **Select map or scene:** Select a web map or scene created by you or shared with you.
- **Add widgets:** Widgets give your app functionality, such as print map and query layers. Each theme has its own preconfigured set of widgets. You can hide or show existing ones and add additional ones. You can also configure some widgets to open automatically as your app starts.
- **Configure attributes:** Attributes allow you to customize your app banner with a logo, title, hyperlinks, and so on.
- **Preview and launch:** Preview the responsive app with popular device screen sizes (2D apps only). When ready, you can launch your app directly, or export your app and deploy it to your own web server.

Pick style	Select map or scene	Add widgets	Configure attributes	Preview and launch

As you configure your web apps interactively, Web AppBuilder generates the configurations in JSON files automatically for you. You do not have to know JSON and the configuration syntax for the widgets themselves. However, if you can edit the JSON files manually, you can have some additional flexibilities in your app configuration.

Widgets

Web AppBuilder provides functions through widgets. Typically, a widget is a JavaScript/HTML component that encapsulates a set of focused functions. Most widgets have a visual user interface.

Web AppBuilder provides more and more widgets with its new releases. As of this writing, Web AppBuilder provides more than 70 core widgets. Based on their relation to map layers, widgets can be categorized into two groups:
- **Data-independent widgets:** For example, Basemap Gallery, Measurement, Draw, and Bookmark widgets are not related to the operational data layers you have in your web map. These widgets need no or little configuration. They are not affected if you switch from one web map to a different web map.

- **Data-dependent widgets:** For example, Query and Chart widgets are related to specific attribute fields of specific layers in your web map. They often require detailed configuration. When you switch from one web map to a different web map, you will need to reconfigure these widgets.

In addition to the out-of-the-box widgets, you can find additional Web AppBuilder widgets from ArcGIS Solutions widgets (**http://arcg.is/2yUXnFC** or **http://solutions.arcgis.com/ shared/help/solutions-webappbuilder-widgets**), which is a set of widgets designed to address specific workflows across industries and from the user community (**http://arcg.is/2jdfu1M** or **https://geonet.esri.com/groups/web-app-builder-custom-widgets**). You can download and deploy these widgets to the following editions of Web AppBuilder:

- **Developer Edition:** Simply copy the widget to the stemapp\widgets folder or stemapp3d\ widgets folder.
- **Portal for ArcGIS (10.5.1 or later):** First, host the custom widget on a web server, and then register the widget URL in Portal for ArcGIS as an application extension (AppBuilder). When you choose widgets to add in Web AppBuilder, you can find the custom widget under the custom tab. For security reasons, only portal administrators can register the custom widget, and public apps will not load the custom widget when anonymous users access it.

Themes

A theme is a template framework representing the look and feel of an app. Content in a theme includes a collection of panels, styles, and layouts, and a set of preconfigured theme widgets. A single app can only use one theme while running.

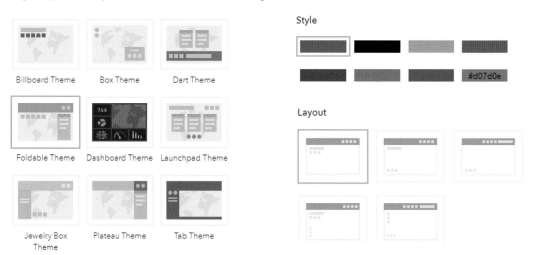

Web AppBuilder for ArcGIS provides many themes, and each theme allows users to choose a color scheme and a layout.

Web AppBuilder for ArcGIS provides the following out-of-the-box themes:

- **Foldable and Tab:** Supports all widget types and can be used for an app with complicated tasks.
- **Dashboard:** All widgets in the panel open simultaneously when the app starts. It is designed to visualize widgets and their communication directly. You can modify the predefined layout by adding or removing grids, or resizing the grids in the panel. Optionally, you can turn on the Header widget to display the logo, the app name, and the links.
- **Billboard:** Designed for apps with simple tasks.
- **Box:** Designed for apps that require a clean look on the map. All on-screen widgets are turned off by default.
- **Dart:** Widgets in the widget controller display like placeholder widgets. You can have multiple widgets open and move them around. All on-screen widgets are turned off by default.
- **Jewelry Box:** Designed for apps with a workflow task. Evolved from the Foldable theme with a focused widget on the left of the app.
- **Launchpad:** Designed for users who like Apple Mac style. Launchpad lets you open multiple widgets, and move, resize, and minimize widgets.
- **Plateau:** Can be used to create a modern and basic application with flat toolbars and widget containers.

Web AppBuilder for ArcGIS Developer Edition

Web AppBuilder Developer Edition provides a great framework for creating new widgets, customizing existing widgets, creating new themes, and building apps with extended functionality. Custom widgets and apps can be shared for free or sold in ArcGIS Marketplace.

In contrast to the embedded editions, you will need to download the Developer Edition, unzip it to a folder on your computer, register its URL, typically as http://[yourmachinename]:3344/webappbuilder, with your ArcGIS Online or Portal for ArcGIS to get an app id, run Web AppBuilder, provide the URL to your ArcGIS Online or Portal for ArcGIS along with this app id, and then you can create web apps in the similar way as the embedded editions of Web AppBuilder. Refer to **http://arcg.is/1C5rYQb** (short for **https://developers.arcgis.com/web-appbuilder/guide/getstarted.htm**) for details about how to get started with the Developer Edition of Web AppBuilder.

The first time you start the Developer Edition of Portal for ArcGIS, you will need to provide the URL to your ArcGIS Online or Portal for ArcGIS and provide an app id. You can get the app id after you register your Web AppBuilder URL in ArcGIS Online or Portal for ArcGIS.

This tutorial

An organization would like to provide a web app that displays data on historic earthquakes and hurricanes to the public.

Data:

A web map is provided to you. This web map uses a map service.

Requirements: The web app should have the following capabilities:

- Zoom to the entire US in its initial view.
- Provide bookmarks so that users can quickly zoom to predefined areas.
- Allow users to print displays as PDFs.
- Allow users to display feature attributes and summaries in charts.
- Allow users to filter features based on their attributes.
- Display the appropriate logo, title, subtitle, and links in the banner.

System requirements:

- A publisher or administrator account in an ArcGIS Online organization.

3.1 Explore the web map

Before you build your app, you will need to familiarize yourself with the web map and the map layers that you will use. A web map has been provided for you in ArcGIS Online. You will explore the web map to understand its layers, fields, and other configurations. Understanding this web map will help you configure the data-dependent widgets later in this section.

1.   Sign in to your ArcGIS Online Organization account.

2.   In the Search box, type **historic disasters GTKWebGIS v3 Sample owner:GTKWebGIS**, and click Search for Maps from the list. On the left of the page, turn off the Only Search in your organization check box.

You should see the search result as illustrated (sample web map for *Getting to Know Web GIS, third edition*).

3.   Click the title of the web map to go to its details page.

4.   On the item details page, under the Layers section, notice that this web map has a dark gray basemap and two operational layers named Earthquakes and Hurricanes.

5.   On the item details page, click the thumbnail image or the Open in Map Viewer button to open the web map in the map viewer.

You should see a time slider appear under the map canvas as well as the earthquakes and hurricanes that occurred within the specified time interval in the time slider.

The time slider indicates that at least one of the layers in the web map is time-enabled; in this web map, both the earthquakes and the hurricanes layers are time-enabled. You can use the Time Slider widget to display the earthquakes and hurricanes by time, which will produce an animation effect. Refer to the Real-time GIS chapter to learn how to enable time on your layers.

6.   Click and drag the right thumb of the time slider to the far right.

January 1, 2000 - March 1, 2009

With the left thumb of the time slider at far left and the right thumb at far right, you will see all the earthquakes and hurricanes in the layers.

7.   In the Details pane, click Legend to see the styles of the layers.

Note that both layers use a color ramp from yellow to red, with yellow indicating lower earthquake magnitudes and hurricane speeds, and red indicating higher earthquake magnitudes and hurricane speeds.

8.   In the map viewer search box, type the name of a hurricane, for example, **Katrina**, and click the Search button or press Enter.

> Katrina   X Q

You will find a section of Hurricane Katrina highlighted on the map.

   Note: The reason you can search hurricanes by name in the search box here is that your web map is configured to support feature search. Feature search is supported by most ArcGIS web clients, including configurable web app templates and ArcGIS API for JavaScript. See the "Questions and answers" section for instructions about how to configure a feature search.

9.   In the Contents pane, click the Content tab, point to the Earthquakes layer, click the Show Table ▦ button, and study the layer attribute fields.

You will use these fields to configure the chart and query widgets in sections 3.4 and 3.5.

If you need to make changes, for example, change the layer styles, remove pop-ups, and configure pop-ups, you can do so now and save it after you are done. Because you are not the owner of this web map, you will need to save it as a new web map if you made any changes.

3.2 Create a web app

1.   Continuing from the last section, click the Share button 🔗.

2.   In the Share window, click Create a Web App.

3.   In the Create a New Web App window, click the Web AppBuilder tab.

Create a New Web App   ×

   Configurable Apps   **Web AppBuilder**

---

4.   Specify your app title, tags, and summary, and click Get Started to open Web AppBuilder for ArcGIS.

Web AppBuilder has two panes—the Design pane on the left and the Preview pane on the right. The Design pane has four tabs: Theme, Map, Widget, and Attribute, which correspond to the four different aspects available to configure your web app.

5.   Click the Theme tab, click through the themes to experiment with them, and see how they look in the Preview pane.

This tutorial will use the Foldable Theme.

6.   Click the Foldable Theme and choose a color style and layout you like.

You can click the Set custom color button to use the color your organization shared, or interactively select a custom color. As you make the changes, look to the Preview pane on the right where you can immediately see how the new setting looks.

7.   Click the Map tab.

The Map tab allows you to choose the web map to use in the app. You started with a suitable web map to use. Otherwise, you can click Choose Web Map to select a different web map.

8. Pan or zoom into the map to cover the earthquakes and hurricanes. Under Set Initial Extent, click Use Current Map View.

Use current map view

Use web map's default extent

9. Click the Attribute tab, set the title to **Historic Earthquakes and Hurricanes**, and in the subtitle, indicate that you designed the app.

Branding

Add logo, title, or subtitle for your app.

| Historic Earthquakes and Hurricanes | A |

| by Peter | A |

Links

+ Add new link

10. Click the logo icon and select the image you would like to use—for example, your organization's logo.

11. Click the Add New Link button to add the URL of your organization or your organization's contact page.

12. At the bottom of the Design pane, click Save.

Launch | Previews | Save

As you complete the rest of the steps in this tutorial, you should save your configuration frequently so that you do not lose your work accidentally.

13. Explore the default widgets in the Preview pane.

Each theme loads with some commonly used widgets by default.

- Click the My Location widget to zoom to your current location.

The ability to find your location is especially useful on mobile devices. This function is available when your app is using HTTPS and the location service is enabled in your browser settings.

- As you zoom and pan across your map, note the Scalebar widget and Coordinate widget showing your current map scale and cursor location.
- In the Search widget, search for an address or place name, for example, Los Angeles, to zoom your map into that location. Or search for a hurricane, for example, Rita.
- Click the Default Extent widget to zoom back to the initial map extent you set.

14. In the lower-right corner (or another corner depending on the layout you chose) of your map, click the Show Map Overview button to bring up the overview window.

15. Click the arrow (now with a reversed direction) to hide the overview window.

16. In the Preview Pane, click the Legend button to see the legend.

You might see the Legend button in the upper-right corner or another location, depending on the theme and layout you chose.

17. Click the Layer List button .

Notice that for each layer, you can click its options button to enable or disable pop-ups, view its attribute table, and view its details.

- At the bottom-center of the map, click the arrow  to bring up the Attribute Table widget. Click the Earthquakes and Hurricanes tabs to see the table for each layer.

The table lists the attributes of the features within the current map extent.

18.   Click the Options button ▦ to see that you can filter the layer, show/hide columns, and select records.

You can allow users to export data to CSVs if you configure the widget to allow so.

19.   Minimize the attribute table.

3.3 Configure data-independent widgets

Data-independent widgets often require little or no configuration. For example, About, Basemap Gallery, Bookmark, Draw, Measure, and Share are such type of widgets.

1. In the Design pane, click the Widget tab.

The Widget tab shows some widgets that are already added to your app, for example, the Attribute Table, Coordinate, and Home Button widgets. Widgets that appear dimmed are turned off but can be turned on.

At the bottom of the list are placeholders for five additional widgets. You will begin by adding several widgets here.

2. Click the first empty widget button. In the Choose Widget window, click Basemap Gallery, and click OK.

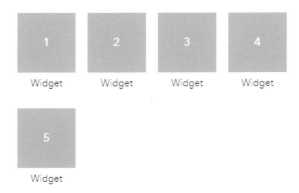

3. In the Configure Basemap Gallery window, click OK.

By default, you are using the basemap gallery setting of your organization. Optionally, you can configure custom basemaps, for example, using your own map services.

Notice that the Basemap Gallery widget has been added to the first widget placeholder in your app.

4. On the Preview pane, click the Basemap Gallery button.

You can choose and switch to a different basemap.

Next, you will add the Bookmark widget.

5. Click the current first empty widget button. In the Choose Widget window, click Bookmark, and click OK.

6. In the Configure Bookmark window, perform the following tasks:

    - Click the Create button.
    - Specify the title as **Western States**.
    - Pan/zoom the map to the western states of the US.
    - Optionally, to specify an icon that represents your bookmark, click the thumbnail.
    - Click OK to add this bookmark.

7. Repeat the previous step to add another bookmark, such as **Southeastern States**.

8. Click OK to close the Configure Bookmark window.

The bookmark widget is added to your app and is ready for use.

9. In the Preview pane of your app, click the Bookmark widget, and then click the bookmarks you defined to see the map extent changes.

📖 **Note:** The bookmarks you defined in the configuration mode are contained in the app configuration and are globally available to all users of your app. You can also add bookmarks in the running mode of the bookmark widget; however, such bookmarks will only live locally in your browser cache, and thus are available only to you.

Next, you will add additional widgets to the header controller.

10. Click the Set the widgets in this controller link.

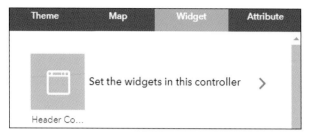

Widgets added to the header controller will appear in the toolbar.

11. Click the Plus button.

12. In the Choose Widget window, click Draw, click Print, and click OK.

You have now added these two widgets to your app.

13. If you need to change the order of your widgets, click a widget button, drag the button to the desired position, and drop it there.

Your two new widgets are data-independent and have default configurations. You can use them right way.

14. In the right corner of the app toolbar, click the Draw button (or the More button, and then the Draw button if your screen size is small). In the Draw window that appears, select a Draw mode.

15. Select a symbol, and experiment with drawing some graphics on your map.

Data-independent widgets may need configuration as well. Next, you will configure the Print widget.

16. In the Design pane, point to the Print widget, and click the pencil 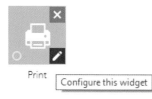.

Print

17. In the Configure Print window, notice that the service URL points to the printing geoprocessing service hosted in ArcGIS Online.

📖 **Note:** If your map has layers from an internal ArcGIS Enterprise, you will need to change the printing service URL to a URL that has network access to your internal server. (See the "Questions and answers" section for details.)

18. Specify the default title as **Historic Earthquakes and Hurricanes**, and the default author as your organization or your name, for instance, and click OK.

19. On your app toolbar, click the Print widget to test how it works, and in the Print window, click Print to print the current map view, including any drawings, to a PDF.

20. When the printing job is done, click the PDF link, examine the PDF, and then close the print window.

The widget added from the widgets collection can be set to open automatically when the app starts. A maximum of two widgets can open automatically: one is on the controller and another is in the placeholder. Next, you will configure the Legend widget to start automatically.

21. In the Design pane, point to the Legend widget, and in the bottom-left corner of the widget, click the dot to change it to dark green.

Legend   Layer List   Draw

You will notice that the legend automatically opens in section 3.6 when you launch the app.

22. In the Design pane, click Save to save your configuration.

3.4 Configure chart-type widgets

In this section, you will enhance your web app by adding a group of chart widgets. Web App-Builder for ArcGIS provides charting capabilities via widgets including the Infographics widget and Chart widget. You will use the former in this section. Chart widgets are data-dependent widgets. You will configure the layers and fields by which these widgets will associate.

1. In the Design pane, click the Plus button. In the Choose Widget window, click Infographic, and click OK.

2. In the Choose a template window, click the thumbnail of the Number template, and click OK.

Choose a template

| Number | Gauge | Vertical Gauge | Horizontal Gauge | Pie | Donut |

3. In the Set data source window, click the Earthquakes layer, and click OK to enter the settings panel of the template.

4. Change the widget title to **Count**. Leave Use selection selected.

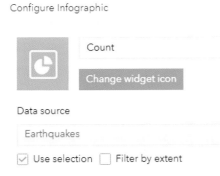

Configure Infographic

Count

Change widget icon

Data source

Earthquakes

☑ Use selection   ☐ Filter by extent

With Use selection selected, the infographic charts will be dynamic to match the earthquake layer selections. You will experiment with it later in this section.

Next, you will modify the layout of the Number widget. Note that the left panel is not only a preview of the graph but also a flexible layout editor.

5.   Click the thumbnails to hide the title and description elements.

6.   Click the image element to highlight it and drag the handle in the upper-right corner to rearrange it to the right side of the number element. Drag and move the divider between the number and the logo to expand the logo element.

7.   Click the logo image to bring up its settings on the right.

8.   Click the Upload button, browse to **C:\EsriPress\GTKWebGIS\chapter3**, select USGS_earthquake_cause.gif, click Open, and click OK.

Image settings

Image   Upload

Alignment

Background color

Link

The image displays the main causes of earthquakes in the US. The widget is configured. Next, you will try it.

9.   In the upper-right corner of the app toolbar, click the Count button to open the widget. Notice that it displays the total number of earthquakes in the layer along with the image you selected.

Next, you will add a pie chart to show the numbers of earthquakes by year.

10.   In the Design pane, click the Plus button. In the Choose Widget window, click Infographic, and click OK.

11.   In the Choose a template window, click the thumbnail of the Pie Chart template, and click OK.

12.   In the Set data source window, click the Earthquakes layer, and click OK.

13.   Change the widget title to **By year**.

14.   Click the thumbnails to hide the title and description elements.

15.   Click the chart area. Under Chart settings, perform the following actions:

- For Display mode, choose Display feature counts by category.
- For Category Field, select Date.
- For Minimum period, select Year.
- Click the Display tab, select Data labels.
- Click OK.

Chart settings		
	Data	Display
Display mode		Display feature counts by categ ▾
Category field		Date ▾
Minimum period		Year ▾
Sort by		⦿ Label [↑][ ]
		○ Value
Maximum categories		100

At this step, you used the Date (earthquake date) field instead of the Year field directly. When you don't have such a year field, or you want to summarize your data by other time units, such as months and days, you can use a date field.

16.   In the upper-right corner of the app toolbar, click the By year button to open the widget. Notice that it displays the count of earthquakes each year in a pie chart.

Next, you will chart the earthquakes' magnitudes.

17. In the Design pane, click the Plus button. In the Choose Widget window, click Infographic, and click OK.

18. In the Choose a template window, click the Column Chart thumbnail, and click OK.

19. In the Set data source window, click the Earthquakes layer, and click OK.

20. Change the widget title to **By magnitude**.

21. Click the thumbnails to hide the title element.

22. Click the chart area. Under Chart settings, perform the following:

   • For Display mode, leave Display values feature by feature as selected.
   • For Label field, select Magnitude.
   • For Value fields, select Magnitude.
   • For Sort by, select Magnitude and click the descending button.
   • Click OK.

For the value fields, you can chart multiple fields, which will create multiple columns per feature.

23. Click the description section. In Text settings, set the Text as **Use the slider to change the magnitude range in the chart**, and click OK.

Text settings	
Text	Use the slider to change the magnitud
Font	Arial
Text size	14
Text color	B I U
Alignment	
Background color	
Link	

Use the slider to change the magnitude range in the chart

24. In the upper-right corner of the app toolbar, click the By magnitude button to open the widget. Explore the widget by doing the following actions:

    • Point to a column to see that the corresponding earthquake is highlighted. You may need to pan the map to include the corresponding earthquake in the map extent.
    • Click and drag the slider under the chart to change the magnitude range.
    • Click and drag an end of the slider to expand or decrease the magnitude range.

When there are too many buttons on the toolbar, you can arrange them into logic groups and make the toolbar cleaner.

25. In the Design pane, click the By year button, drag and drop it to the Count button.

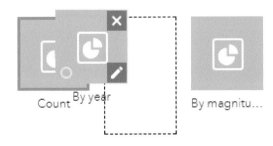

26.   Notice these two widgets became a group.

27.   Click the By magnitude button, drag and drop it to the group.

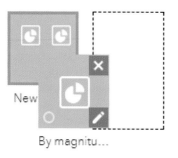

By magnitu…

28.   Point to the New Group and click the Configure this widget button ✎. In the New Group window, change the Label to **Earthquake Charts**, and click OK.

29.   In the upper-right corner of the app toolbar, click the Earthquake Charts button to open the charts. Notice the three charts appear in one panel.

30.   In the Design pane, click Save to save your configuration.

3.5 Configure filter-type widgets

In this section, you will further enhance your app by adding filter or query capabilities. Web App-Builder provides many filter or query types of widgets to perform attribute, spatial, or both query capabilities. Spatial queries can use the current map extent, a point or shape that users will draw, or features from a different layer. Filter and query types of widgets are data-dependent widgets, so you will need to configure their associated layers and fields.

In this section, you will first add the Info Summary widget, and then add a Filter widget. The Info Summary widget queries your layers using the current map extent, and lists the features contained in the extent.

1.   Continuing from the last section, click the Plus button to add additional widgets to the header controller.

2. In the Choose Widget window, click Info Summary widget, and click OK.

3. In the Configure Info Summary window, click Add Layer. Make sure the Earthquakes layer is added. Point to the Earthquakes layer. Under Display Options, click the Configure this widget button ![icon].

Configure Info Summary   ✕

Info Summary

Change widget icon   Learn more about this widget

⊕ Add Layer

Layer	Label	Display Options	Refresh	Actions
Earthquakes ▾		✎ ●	☐	⬆⬇✖

—Options——————————————————————————————— [Edit]

4. In the Display Settings window, click the Panel tab, and complete the following actions:

   • Set the three fields to be **Magnitude**, **Location**, and **Year**.
   • Set the label for Magnitude as **Magnitude**.
   • Leave the labels for the other fields blank.
   • Click OK.

Display Settings   ✕

      Symbol   Panel

Layer Options

⊕ Add Field

Field	Label	Actions
Magnitude ▾	Magnitude	
LOCATION ▾		
YEAR ▾		

Group Options

☐ Group features

[OK] [Cancel]

5. Under Options, select Display Feature Counts, and then click OK.

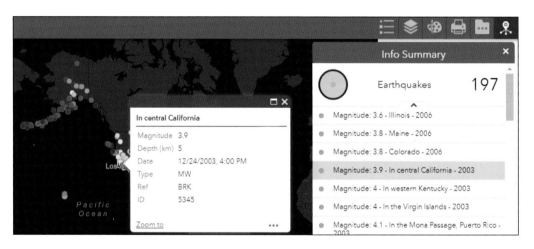

Next, you will try the widget to see how it behaves.

6. In the upper-right corner of the app toolbar, click the Info Summary widget to open it. Click the arrow under the layer to expand the feature list. Pan or zoom the map to see the feature count and list update.

For each feature in the list, the attribute fields you configured previously are displayed.

7. In the Info Summary widget, click a feature in the list and the pop-up of the feature will display on the map.

Similarly, you can add the Hurricanes layer into the widget. If you do so, in the Display Settings of the Hurricanes layer, you should group the features by the Name field. You will group the features because in the data, each hurricane path is broken into multiple lines corresponding to the hurricane wind speed changes. The group option will make it easier for your users to study each hurricane.

8. In the Design pane, click Save to save your configuration.

Next, you will configure the Filter widget.

All widgets in the header controller or header toolbar share the same panel, which means only one widget displays at a time. As one displays, the previous one closes. For your users to see both the Filter widget and the Charts widget group at the same time, you will add the Filter widget outside of the Header Controller.

9. In the Design pane, click the left arrow in the Widgets tab to navigate out of the header controller.

← Widgets

Set widgets managed by Header Controller

10. Click the first empty widget button. In the Choose Widget window, click Filter, and click OK.

11. In the Configure Filter window, click New Filter.

Configure Filter   ×

▼   Filter

      Change widget icon   Learn more about this widget

Add one or more filters to the widget and configure parameters for each of them.

  + New filter   Layer

                         Title   ⊡ Earthquakes
                                    ⊟ Hurricanes
                         Icon   ▼ ▾

12. In the Layer drop-down list, choose Earthquakes.

Next, you will add a filter expression to filter the layer by earthquake magnitude.

13. Click the Expressions tab, click Add expression, and perform the following tasks:

- In the field list, click **Magnitude (Number)**.
- In the operator list, click and choose "is at least."
- Select Ask for values.
- Leave Prompt as "Magnitude is at least."

Info	Expressions	Options

+ Add expression   + Add set

Get features in the layer that match [All ▼] of the following expressions

| Magnitude (Numbe ▾ | is at least ▾ | | ⚙ ✕ |

☑ Ask for values

Prompt   Magnitude is at least

Hint

Next, you will add another filter expression to filter the layer by earthquake depth.

14. Click Add expression again, and perform the following tasks for the new expression:

- Notice "Get features in the layer that match All of the expressions," which essentially means both this expression and the previous expression will have to be met.
- In the field list, click Depth (km) (Number).
- In the operator list, click and choose "is at most."
- Select Ask for values.
- Leave Prompt as Depth (km) is at most.

| Depth (km) (Numbe ▾ | is at most ▾ | | ⚙ |

☑ Ask for values

Prompt   Depth (km) is at most

Next, you will add one more expression to filter the layer by earthquake year.

15. Click Add expression again, and perform the following tasks for the new expression:

- In the field list, click Year (Number).
- In the operator list, click and choose "is."
- Click the Set input type button ⚙ and select Unique.
- Select Ask for values.

- Leave Prompt as Year is.
- Leave List values as Values filtered by previous expressions.
- Click OK.

| Year (Number) | ▾ | is | ▾ | | ▾ | ⚙ | ✕ |

☑ Ask for values

				Value
Prompt	Year is			Field
Hint				Unique	✓

List values: Values filtered by previous expressions   ▾

Next, you will explore this widget to understand what you have configured.

16. In the Preview pane, click the Earthquake Charts button and click the Filter button. Notice both the Charts and the Filter widgets are open.

17. In the Filter widget, click the Year is list and notice it has nine options, from 2000 to 2008.

18. Specify magnitude is at least 6.5, and depth is at most 10 km.

19. Click the Year is list, and notice it has fewer options. Understand that its values have been filtered by the magnitude and depth expressions.

20. Choose Year is 2007, and click the radio button in the upper-right corner of the Filter widget to apply the filters.

21. Notice how the map displays only the earthquakes matching the filters. Also notice how the charts have been updated to display the result of the filters.

22. Click the radio button in the upper-right corner of the Filter widget again to remove the filter. Notice the map and the charts have been updated to display all earthquakes.

23. In the Design pane, click Save to save your configuration.

3.6 Preview and share your app

In the previous steps, you saw the effects of your configurations in the Preview pane. You can further preview your app in various mobile devices and in its own browser window.

1. At the bottom of the Design pane, click Previews.

You will see a list of popular mobile devices.

2. Choose the type of device you wish to preview, or specify a custom screen resolution.

3. In the Preview window, try each of the widget buttons and see how your app works on different mobile devices.

4. In the upper-right corner of your browser, click the Phone Orientation button ⬚ or ▭ to change the orientation of the device. Try your app to see how it behaves in the new orientation.

If your smartphone or tablet has a QR scanner app, you can scan the QR code in the lower-left corner area to view your app on your mobile device directly.

5. In the Design pane, click < Configure to return to the configuration mode.

6. Click Launch to view your app in the full web browser.

Your app will display in a new web browser or tab. The URL is your app URL that you will share with your audience and your instructor.

7. Go to your ArcGIS Online content list, and find and select the app.

8. Click the Share button to share the app with Everyone, and click OK.

In this tutorial, you created a web app that provides a set of functions without programming. You chose the necessary widgets, configured them, and combined them to create a useful and usable web app. The widgets you configured are among the widgets most commonly used. You also can try other widgets and explore how they work.

QUESTIONS AND ANSWERS

1. My Print widget does not work. Instead, an error message appears that reads, "Error. Try again." Why?

   Answer: A common reason for this error message is that your web map contains a layer from an internal ArcGIS Enterprise, which is inside your network firewall. The default printing service configured in your Print widget often comes from ArcGIS Online. ArcGIS Online printing service sits outside your network firewall and typically cannot access your internal server. As a result, the printing service cannot ask your ArcGIS Enterprise to generate a map or return the data.
   To fix this problem, replace the default printing service URL with an internal printing service URL. Your ArcGIS Enterprise comes with a built-in printing service, which you can find in the Utilities folder of your ArcGIS for Server Services Directory. Some organizations choose to stop this printing service. You can ask your GIS admin to start the service using ArcGIS for Server Manager (navigate to **http://your_ArcGIS_Server_name:6443/arcgis/manager**).

2. In the tutorial, I searched for hurricanes directly in the place/address search box and in Web AppBuilder Search widget. How can I configure my web map to support this?

   Answer: This capability is called feature search, which allows users to locate features in the same search box as the address and place name search. For example, enabling search on your parcel layer would allow users to find specific parcels simply by entering a parcel ID in the search box. For your users, this way to locate features is consistent with the way they locate an address or place name.
   To configure feature search, you can go to the item details page of your web map and click Settings. In Application Settings and under Find Locations, you can enable By Layer, select the Layer and Field you want to allow your users to search, and choose a condition for comparison. You can also configure the Hint text, which will appear in the search text box and tell users for what they can search.

Application Settings

Select the tools and capabilities to enable in applications that access this web map.

☑ Routing
☑ Measure Tool
☑ Basemap Selector
☑ Find Locations [-]

    Hint text

    | Place, address, or hurricane name |

    ☑ By Layer

| Earthquakes ▼ | LOCATION ▼ | Contains ▼ | ✕ |
| Hurricanes ▼ | Name ▼ | Contains ▼ | ✕ |

    ☑ By Address

3. Can Web AppBuilder for ArcGIS work directly with web services?

Answer: Yes and no.

Web AppBuilder is not designed to work directly with web services when specifying a web app's map content. This workflow (adding services directly into the map) is not supported and is not recommended. Instead, the Web AppBuilder works with web maps and web scenes, which can encapsulate web services. Authoring a web map is easy and empowers non-GIS experts to make their own maps. Web AppBuilder is designed to extend the reach of non-GIS experts so they can easily create a custom web app for their web maps/scenes. This design strengthens the concept of extending the power of GIS and spatial mapping throughout an organization.

However, Web AppBuilder does work directly with web services via its many widgets. You can configure widgets to work directly with your web services by specifying their REST URLs when you set the properties of a widget. For example, you can specify your own basemap services for the basemap gallery widget, a custom geoprocessing service for the geoprocessing widget, or your own layer as the data source for the add data, query, chart and other widgets.

4. How can I have multiple widgets open at the same time?

Answer: You can use the approach in section 3.5, which can display an in-panel widget and an off-panel widget at the same time. You can also use the Dashboard theme, which allows you to display 4, 6, or 9 widgets at the same time.

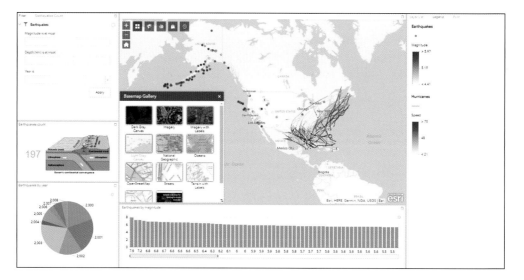

5. A layer in my web map doesn't allow me to use it with the query widget. Why?

Answer: Some widgets require certain layer types. The query widget requires a web service layer. Those data you added into your web map directly cannot support the query widget. Try publishing the data as a feature layer or other web service types, and then add the layer to your web map.

ASSIGNMENT

Assignment 3: Build a web app using Web AppBuilder for ArcGIS.

Data:

No data have been provided. Instead, create a web map using the feature layer you published before, or find a web map in ArcGIS Online.

Requirements:

- The initial map extent should zoom to your study area.
- Provide bookmarks that allow users to quickly zoom to predefined areas.
- Let users print their maps as PDFs.
- Enable users to chart selected features with selected attributes.
- Allow users to filter or query for features by specifying values for multiple attributes.

What to submit:

- The URL to your web app.

Resources

"Configuring Web Apps Using Web AppBuilder for ArcGIS," https://www.esri.com/training/catalog/57f45 05e362fd58367ab5302/configuring-web-apps-using-web-appbuilder-for-arcgis (or http://arcg.is/ 2hBpKl1).

"OSO Mudslide—Before and After," http://learn.arcgis.com/en/projects/oso-mudslide-before-and-after (or http://arcg.is/1MxzBmA).

"Web AppBuilder for ArcGIS online help document site," http://doc.arcgis.com/en/web-appbuilder (or http://arcg.is/1DAV2Kb).

"Web AppBuilder for ArcGIS widgets overview," http://doc.arcgis.com/en/web-appbuilder/create-apps/ widget-overview.htm (or http://arcg.is/2nywVgy).

"Web AppBuilder for ArcGIS: An Introduction," Jianxia Song and Derek Law, https://www.youtube.com/ watch?v=nIYE-_Nhdec (or http://bit.ly/2zevTvw).

"Web AppBuilder for ArcGIS: Customizing and Extending," Moxie Zhang and Gavin Rehkemper, https:// www.youtube.com/watch?v=9JttgbuZsEs (or http://bit.ly/2yYLbDB).

"What's new in Web AppBuilder for ArcGIS," https://www.esri.com/search?filter=Blogs&q=What%E2%80 %99s+New+in+Web+AppBuilder+for+ArcGIS&search=Search (or http://arcg.is/2yceSSC).

CHAPTER 4

Mobile GIS

We live in the post-PC era. We have far more smartphones and tablets than desktops and notebooks. Today, we access the web primarily from mobile devices. With their advantages in mobility and location awareness, mobile devices are becoming the pervasive client platform for Web GIS. Mobile GIS is an indispensable part of our lives and our enterprise operations. This chapter starts with an overview of mobile GIS and its related frontiers of application, including location-based services (LBS), volunteered geographic information (VGI), virtual reality (VR), and augmented reality (AR). The chapter then introduces editable feature layers (the server-side support for data collection), the three options for building apps (browser-based, native-based, and hybrid-based), and Esri mobile apps, including Collector for ArcGIS®, Survey123 for ArcGIS®, Workforce for ArcGIS®, and AppStudio for ArcGIS®. The tutorial sections teach you how to create an editable feature layer and use it with Collector for ArcGIS, how to design a smart survey and use it with Survey123, and how to create a native app using AppStudio for ArcGIS.

Learning objectives

- *Understand the concepts of mobile GIS, LBS, VGI, VR, and AR.*
- *Understand the three approaches to building mobile apps.*
- *Configure editable feature layers and feature templates.*
- *Use Collector for ArcGIS to collect GIS data.*
- *Create smart surveys using Survey123 web designer.*
- *Collect data using Survey123.*
- *Use AppStudio for ArcGIS to build native apps.*

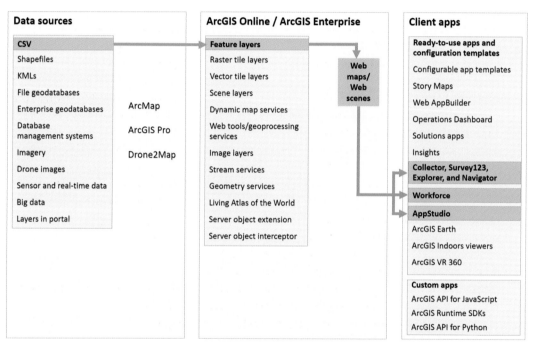

ArcGIS offers many ways to build web apps. The green lines in the figure highlight the technology presented in this chapter.

Mobile GIS concepts, advantages, and frontiers of application

Mobile GIS refers to GIS for use on mobile devices and has the following advantages over traditional desktop GIS:

- **Mobility:** Mobile devices are not hindered by wires or cables. These devices extend GIS to areas where wiring is infeasible or costly and users most need GIS.
- **Location awareness:** You can use GPS, cellular networks, Wi-Fi networks, Bluetooth technology, and other technologies to pinpoint the current location of a mobile device. You also can use the compass, gyroscope, and motion sensors of a mobile device to determine the device's direction, tilt angle, and moving speed.
- **Ease of data collection:** Mobile GIS can replace existing paper-based workflows, which are prone to errors that arise when surveyors draw on the map in the field and enter data manually back in the office. Mobile GIS can replace such paper datasheets, reduce costs, and improve the accuracy of your data.
- **Near real-time information:** The live connection of mobile networks greatly enhances the temporal dimension of GIS. This dimension gives mobile GIS the potential ability to monitor the spatial and temporal aspects of the world around us.
- **Large volume of users:** Mobile devices create a pervasive platform for GIS and deliver GIS to the hands of billions of people.
- **Versatile means of communication:** Integrating with voice, short message, photo, video, email, and many social networking apps, mobile devices facilitate collaboration and communication among professionals and consumers.

Mobile GIS is built on top of the following technologies (among others):

- **Mobile devices:** Mainly smartphones and tablets
- **Mobile operating systems:** Mainly Android and iOS
- **Wireless communication technology:** Including Bluetooth, Wi-Fi, and cellular network technology, which has evolved into 4G (fourth-generation) and 5G
- **Positioning technology:** Including the navigation satellite-based approach such as GPS, cellular network-based approach, Wi-Fi-based approach, and various indoor positioning technologies that are evolving

Mobility comes at a price. The small size of mobile devices imposes limitations on CPU speed, memory size, battery power, bandwidth and network connections, screen size, and keyboard size. Some of these limitations are becoming less severe as the technology advances, but they still must be considered when designing and developing mobile applications.

The mobile platform presents a new paradigm for GIS. Mobile GIS has broad applications for consumers and organizations. Individual consumers use mobile GIS to learn what is nearby, where to eat, how to get there, and share what they saw. Organizations need mobile GIS to complete tasks such as field mapping, data query, field inspection and inventory of assets, tracking assets and field crews, field surveys, incident reporting, and parcel delivery.

Mobile computing is integrating into everyday life, and 5G, the next generation of the wireless communications network, is coming. As GIS is integrated with a multitude of computing devices and connected to the web with the faster 5G network, GIS will become ubiquitous.

Popular types of apps and frontiers

Mobile GIS is related to many popular types of apps and frontiers, including LBS, VGI, VR, and AR.

- LBS are services offered through a mobile device and consider the device's geographical location. LBS typically provide information or entertainment. Desktop GIS apps allow you to click a POI to get its information. With LBS, you essentially become the mouse cursor on a map of 1:1 scale—the real world. As you enter an area or get close to a POI, the mobile apps on your phone know where you are and push information to you about this area or POI.

- VGI is the digital spatial data produced voluntarily by citizens rather than by formal institutional data producers. The term "VGI" was coined in 2007 to refer to user-generated content in the geospatial field. VGI is often reported using mobile devices. Various navigation and traffic apps, for example, collect VGI about highway traffic, accidents, and police car locations that mobile users contribute voluntarily. Other examples of VGI include georeferenced tweets collected by Twitter and georeferenced photos uploaded to Flickr and Instagram. VGI marks a research frontier of significant practical value. With the help of crowdsourcing, in which large numbers of citizens act as sensors, VGI can enhance early warning systems for natural disasters, epidemic, or real-time social events monitoring. VGI is also an important source of big data and supports tremendous new business opportunities.

- VR is the computer-generated simulation of a 3D map or environment that can be interacted with in a seemingly real or physical way by a person using special electronic equipment, such as a helmet with a screen inside or gloves fitted with sensors. Today, smartphones and VR glasses, such as VR cardboards, provide an inexpensive platform that makes the fun, immersive experience of VR available to everyone. Refer to chapter 7, "3D Web Scenes," for more details.

- AR is an enhanced version of reality created using technology to overlay digital information on the view through a device, such as a smartphone or tablet camera. AR is often related with mobile GIS because a mobile device can retrieve information based on your location, the direction you are facing, the tilt angle of your camera, the live view in your camera, and can overlay the retrieved information in your camera view. Refer to chapter 7 for more details. Here are two examples of AR apps:
    - An AR-based travel app can superimpose a building's historic pictures on its current view through your phone's camera. If you point your camera to a restaurant, the app can display services, reviews, and open hours of the restaurant in a pop-up window on your phone.
    - An AR-based pipeline map that can retrieve the underground pipeline information as you hold your phone downward. The app can overlay the pipelines with what you see through your camera view and help you "see" underground.

Editable feature layers

You have published feature layers in the previous chapters. These feature layers are read only and can support mapping and search capabilities for mobile GIS. Data collection and VGI types of applications essentially involve adding, removing, and updating geometries and attributes. To support such web-based editing capabilities in Web GIS, you will need to enable editing on your feature layers and share the layers with your users.

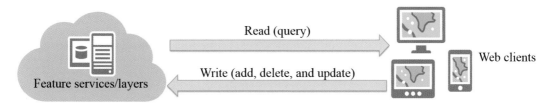

With editing enabled, feature layers or services can provide web clients with read and write access to the server data.

> 📝 **Note:** ArcGIS Online and ArcGIS Enterprise provide web editing capabilities to support simple feature editing. ArcGIS Pro and ArcMap are the best options for more complex editing operations such as topologies and geometric networks.

As the owner of a hosted feature layer or an administrator in the organization, you can choose from the following options to control what edits can be made to your hosted feature layer, if any:

- Add, update, and delete features.
- Add and update features.
- Add features. For instance, you allow the public to report new incidents but don't allow them to delete the information reported previously.
- Update features.
- Update attributes only. For instance, this option is good for observation towers whose location shouldn't be allowed to change but whose attributes may be updated.

If you enable attachments on your feature layer, this setting will allow your users to attach photos, videos, and other files with the features. Users can attach multiple files for one feature, and each file can be up to 10 MB.

You can also enable editor tracking. Editor tracking refers to the ability to track who has changed the data of a feature layer, and when. Editor tracking can help create more accountability and quality control over the edited data. Tracking can also support ownership-based access control, which allows you to limit access so only the user who created a certain feature can access that feature.

Feature layer views

If you have publishing privileges and need a different view of your hosted feature layer beyond changing its style and presentation—for example, you want to apply different editing capabilities or share the data with different groups—you can create a view from your hosted feature layer. Many organizations need to share data with the public and simultaneously allow members within the organization to keep that data up to date. Hosted feature layer views provide a direct way to do this. When you publish your hosted feature layer, you can share it with certain members of your organization who need to edit it. Then, for the general public, you can create a hosted feature layer view that references the original hosted feature layer but with editing capabilities disabled. Because the two layers share the same data, as members edit the original hosted feature layer, the general public will see those changes immediately.

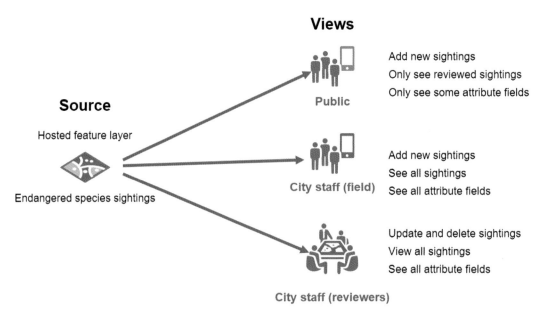

A view and its source feature layer reference the same data. You can independently set how the view is shared with others, how it's drawn, what features are displayed (filtering), and whether the layer can be edited.

You can create views on the item details page of your hosted feature layer. When you create smart surveys using Survey123, ArcGIS will create a feature layer and the necessary views for you behind the scenes automatically.

Feature template

A feature layer can contain a feature template, which you can define in ArcGIS Desktop or portal map viewer. A feature template defines the types of data items that users can add to a layer. A template ensures data integrity and makes editing easier for your end users.

- **Data integrity:** A feature template for a layer that represents schools, for example, might allow editors to classify a new feature as an elementary, middle, or high school. In a school feature template, you might preset these three options so that users can choose only one of these three types of schools. This prevents users from entering invalid values for such an important attribute.
- **Ease of editing:** A feature template can have preset symbols and default values for one or multiple fields. Preset symbols make it easier for users to know what type of feature they are using or adding. With the default attribute values, users do not have to type these values manually, which is particularly convenient for mobile users.

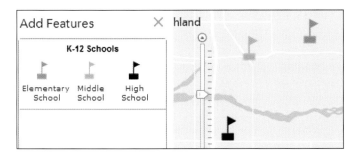

The feature template for K–12 schools defines the types of schools that users can add.

Mobile app development strategies

Choosing a mobile app development strategy depends on the development team's skillset, the application's required functionality, the targeted platform(s), and the amount of funding available. Mobile application development includes the following approaches:

- **Browser-based approach:** This approach builds apps using HTML, JavaScript, and Cascading Style Sheets (CSS). Users access these apps via mobile web browsers. This strategy can potentially reach all mobile platforms. Browser-based apps typically are less costly and quicker to develop than native apps. However, browser-based apps can access only a limited amount of a device's native features. As such, the user experience and performance of browser-based apps typically cannot compete with the experience of using native apps, which don't have the limitation of web browsers.
- **Native-based approach:** The apps you download and install on your mobile device are native apps. The native-based approach requires native development skills, such as

Objective-C or Swift for iOS, Java for Android, and .NET for Windows Phone. These apps typically have deep-level access to device hardware and other resources, and typically have better performance than browser-based apps. However, native apps are often more expensive to develop than JavaScript apps, and one app cannot run on multiple platforms.

- **Hybrid-based approach:** This approach integrates native components and HTML/JavaScript/CSS to build native applications. You can achieve this in many ways. The simplest way embeds a web control into a native app to load HTML and JavaScript contents. More advanced methods include the use of frameworks such as Adobe PhoneGap to allow for deeper integration with the native platform.

ArcGIS offers APIs that work for each of these approaches. ArcGIS API for JavaScript supports both browser-based and hybrid-based approaches. ArcGIS® Runtime SDK (Software Development Kits) for iOS, ArcGIS® Runtime SDK for Android™, ArcGIS® Runtime SDK for Windows Phone™, and ArcGIS® Runtime SDK for Qt support the native-based approach. Each of these APIs ultimately offers similar core functionalities that include editing, layers, graphics, and geometry, as well as accessing ArcGIS web maps and web services and a variety of tasks such as querying, identifying, searching, and geoprocessing.

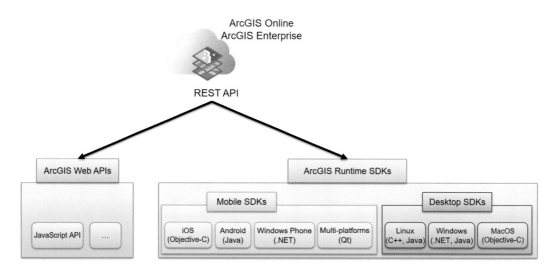

ArcGIS provides a suite of web APIs and runtime SDKs, which interact with ArcGIS Online, Portal for ArcGIS, and ArcGIS for Server via ArcGIS REST API.

Because ArcGIS Web APIs, Runtime SDKs for Mobile, and Runtime SDKs for Desktop share similar concepts, understanding one helps you learn another. For example, you may refer to a sample code of ArcGIS API for JavaScript when you develop Android apps using Java.

ArcGIS native apps

ArcGIS provides a suite of native apps, including Collector for ArcGIS, Survey123 for ArcGIS, Workforce for ArcGIS, Navigator for ArcGIS®, and Explorer for ArcGIS®.

Collector for ArcGIS

Collector can perform the following functions:

- Collect and update data using the map or GPS.
- Download maps to your device and work offline.
- Collect points, lines, areas, and related data.
- Fill out easy-to-use, map-driven forms.
- Attach photos to your features.
- Use professional-grade GPS receivers.
- Search for places and features.
- Support smarter forms and provide enhanced high-accuracy 3D data collection capabilities in the upcoming versions of Collector.

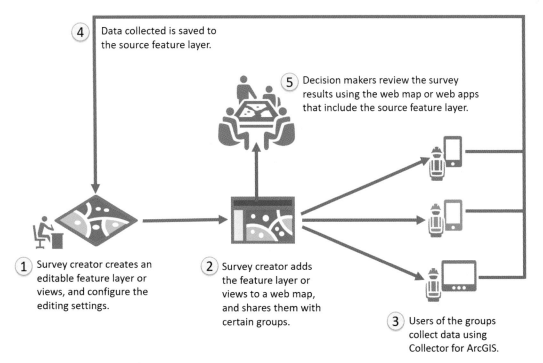

Workflow to use Collector for ArcGIS.

Survey123 for ArcGIS

Survey123 provides the following capabilities:

- Design surveys with predefined questions that support domains and feature templates, default values, embedded audio and images, and rules (for example, if the answer to one question is yes, then show a related question; otherwise, do not show the related question).
- Capture field data using an intuitive form-centric data-gathering solution.
- Store survey results in hosted feature layers that you can easily share with organization users.
- Perform online and offline data collection.

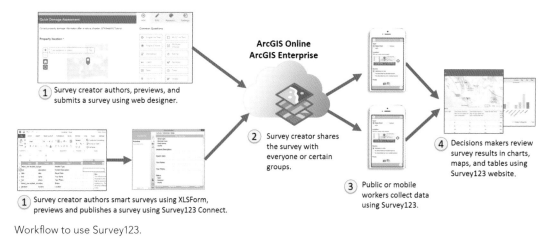

Workflow to use Survey123.

Workforce for ArcGIS

Workforce for ArcGIS is a mobile solution that allows for better field workforce coordination and teamwork. This solution includes web apps and a mobile app.

- **Project owner** can use the Workforce website to create and configure Workforce projects, define assignment types, and assign users dispatcher and mobile worker roles.
- **Dispatcher** can use the Workforce website to create assignments, prioritize them, and assign them to mobile workers, notifying them of their tasks and deadlines. Dispatchers can also track the mobile workers and view their locations on the map.
- **Mobile workers** use the Workforce mobile app. They will get notified as assignments come in. They can easily view and process work assignments, provide updates on work status, submit photos or other attachments, inform others of their location, and use integrated mobile apps such as Collector and Survey123.

Workflow to use Workforce.

Explorer for ArcGIS and Navigator for ArcGIS

Explorer for ArcGIS allows users to search and display web maps, search for places and features in web maps, share maps with other users as images and links, and give presentations with inter-active maps.

Navigator for ArcGIS is more than a consumer-level navigation app. In addition to common functions such as searching for locations, getting directions, and voice-guided, turn-by-turn navigation, Navigator for ArcGIS allows you to use your organization's road network data and access downloadable maps for offline navigation.

AppStudio for ArcGIS

AppStudio for ArcGIS provides a template-based approach for building cross-platform apps with-out coding. You select an out-of-the-box template, configure it by following an easy-to-use wizard, link it to an existing web map/app or feature layer, specify your own branding, and build the app for one or multiple platforms, including iOS, Android, Windows, OS X, and Linux. AppStudio lets developers and organizations quickly turn their existing web maps/apps/layers into beautiful consumer-friendly, cross-platform native apps.

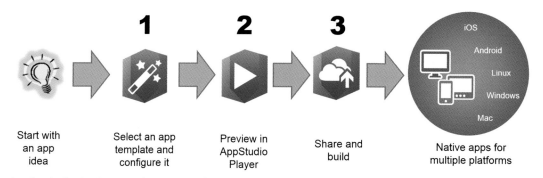

1	**2**	**3**	
Start with an app idea	Select an app template and configure it	Preview in AppStudio Player	Share and build	Native apps for multiple platforms

AppStudio for ArcGIS provides app templates. You can select one template, configure it, preview it, and build native apps for your target platform(s).

A major use of mobile GIS is data collection. ArcGIS provides many solutions for this purpose. Browser-based apps such as the Survey123 browser app and native app, Story Map Crowdsource, and GeoForm template are easier to use and allow anonymous access, and thus are good for crowdsourcing and VGI types of apps. Collector for ArcGIS requires an account in ArcGIS Online or Portal for ArcGIS, and is more for organizational uses. AppStudio provides customization capabilities but requires deployment.

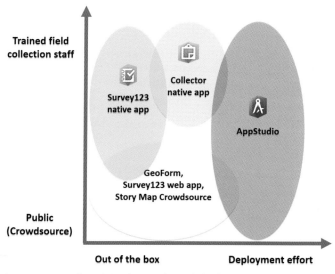

A comparison of ArcGIS solutions for mobile data collection.

This tutorial

This tutorial includes multiple exercises.

- **Exercise 1:** A city would like to have the public and staff of the public works department report non-emergency incidents.
  - Section 4.1 prepares an editable layer and a web map for use with Collector for ArcGIS.
  - Section 4.2 illustrates how city staff report incidents using Collector for ArcGIS and how the public reports incidents using a browser-based app.
- **Exercise 2:** After a hurricane, most houses in a city were affected. The city doesn't have enough assessors, and thus would like citizens to conduct the initial damage assessment.
  - Section 4.3 designs a smart form for Survey123.
  - Section 4.4 collects data using the Survey123 mobile app.
- **Exercise 3:** Create a new native app for field data collection.
  - Section 4.5 creates a native app using AppStudio for ArcGIS.
  - Section 4.6 deploys the native app to your device.

System requirements:

- An ArcGIS Online publisher or administrator account
- Collector for ArcGIS mobile app
- Survey123 mobile app
- An iOS or Android smartphone or tablet (for sections 4.2 and 4.4)
- An Android device or a desktop/laptop computer (for section 4.6)
  - Section 4.6 is about how to deploy native apps to devices. Deploying apps to an iOS device is quite involved and requires your being a registered iOS developer with an annual fee to Apple, your certificate and provisioning file. Deploying apps to an Android device or a desktop/laptop computer is much easier. Therefore, section 4.6 illustrates the procedure using an Android device, or a computer in case you don't have an Android device.

4.1 Create an editable feature layer

Collector for ArcGIS requires a web map that includes an editable feature layer. The web map must be yours or shared with a group of which you are a member. You previously learned how to publish feature layers and create web maps. Thus, this section will be brief.

1. Open C:\EsriPress\GTKWebGIS\Chapter4\311_Incidents.csv in a text editor or Microsoft Excel, and study its data fields.

Incident_Type	Description	Longitude	Latitude	Date	Reported By	Contact	Status
Pothole	A deep one!	-118.2348328	34.0523553	10/21/2017	Peter	peter@peter321.com	Open
Street Sign	Fell off	-118.2678776	34.04047861	10/22/2017	Peter	peter@peter321.com	Open
Street Light	Light is off	-118.2232456	34.04268339	10/23/2017	Peter	peter@peter321.com	Open
Manhole Cover	Missing	-118.2703881	34.05512867	10/24/2017	Tim	909-1234571	Open
Dead Animal	In the center lane	-118.2388024	34.02804896	10/25/2017	Tim	909-1234571	Open
Graffiti	On the wall	-118.293131	34.052439	10/26/2017	Tim	909-1234572	Open

The incidents already in the sample CSV file can be actual incidents or merely placeholders that can be deleted once you have created your feature layer. These existing records serve two purposes:

- **Help define attribute field types:** Unlike databases and shapefiles, which explicitly define attribute field types, CSV files can't do this task. ArcGIS must determine the field types in CSV files based on their values. For example, if a field contains the value "10/21/2013," ArcGIS will recognize the field as a date value.
- **Help define attribute domain values:** The Incident_Type field has a list of preset values. You will use these values to create a feature template, and thus limit the allowable values for this field. You can define feature templates other ways; however, it is easier to create a feature template based on all the possible values in a field of a CSV file.

2. Close Excel or the text editor. Do not save the CSV.

3. Open a web browser and go to ArcGIS Online (**http://www.arcgis.com**) or your Portal for ArcGIS. Sign in with a publisher or administrator account.

4. Click Content on the main menu bar, under My Content, click Add Item ✛, and then click From my computer.

5. In the Item from my computer window, perform the following tasks:

- For File, browse to C:\EsriPress\GTKWebGIS\Chapter4\311_Incidents.csv, and select it.
- For Title, use the default, or specify a new one.
- For Tags, specify **311**, **Incidents**, **VGI**, and **GTKWebGIS**.

- Make sure the check box next to Publish this file as a hosted layer remains selected.
- For Time Zone, select your time zone.
- Click Add Item.

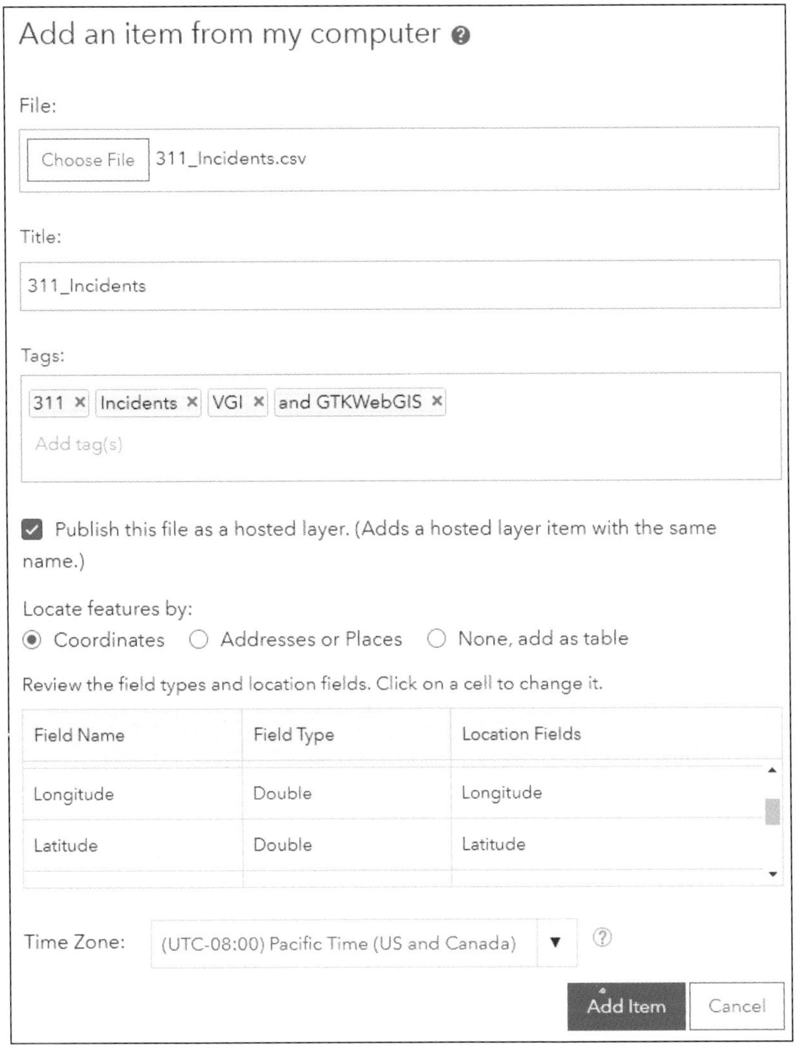

Add an item from my computer ❷

File:

| Choose File | 311_Incidents.csv |

Title:

311_Incidents

Tags:

311 ✕ Incidents ✕ VGI ✕ and GTKWebGIS ✕

Add tag(s)

☑ Publish this file as a hosted layer. (Adds a hosted layer item with the same name.)

Locate features by:
◉ Coordinates   ○ Addresses or Places   ○ None, add as table

Review the field types and location fields. Click on a cell to change it.

Field Name	Field Type	Location Fields
Longitude	Double	Longitude
Latitude	Double	Latitude

Time Zone: (UTC-08:00) Pacific Time (US and Canada) ▼   ⑦

Add Item   Cancel

The item details page appears as your CSV file is being published as a hosted feature layer.

6. On the item details page, look for the Layers section. Under the layer 311_ Incidents, click Enable Attachments.

Next, you will delete the latitude and longitude fields. They were needed for creating the feature layer but are no longer needed after your feature layer is created. The geometries of your points are managed by the feature layer itself internally. Your end users will not need to fill in the latitude and longitude fields.

7. Click the Data tab.

8. In the table header, click the Longitude field, and choose Delete. When prompted, confirm the deletion.

Overview	Data	Visualization	Usage	Settings

Double-click a value in the table to change it.

311_Incidents (Features: 6, Selected: 0)

Incident_Type	Description	Longitude		Latitude
Pothole	A deep one!	↑≣ Sort Ascending		5235
Street Sign	Fell off			4047
Street Light	Light is off	↓≣ Sort Descending		4268
Manhole Cover	Missing	Q⊟ Show Detailed View		5512
Dead Animal	In the center lane	▦ Calculate		2804
Graffiti	On the wall	🗑 Delete		5243

9. Repeat the previous step to delete the Latitude field.

10. While still on the item details page, click the Settings tab.

11. Click Set Extent.

12.   In the Set Extent window, zoom and pan the map to your city or the area you
      expect to collect data, Click the Draw Extent button, draw an extent on the
      map to define your layer extent, and then click OK.

13.   Under Editing, select Enable editing, and leave Add, update, and delete
      features as selected.

Editing

☑ Enable editing.
☐ Keep track of created and updated features.
☐ Keep track of who created and last updated features.
☐ Enable Sync (disconnected editing with synchronization).

• Who can edit features?
  Share the layer to specific groups of people, the organization or publicly via the Share button on the Overview tab. This layer is
  currently shared with: Everyone (public)

• What kind of editing is allowed?
  ◉ Add, update, and delete features
  ○ Add and update features
  ○ Add features
  ○ Update features
  ○ Update attributes only

14. At the bottom of the page click Save.

15. Click the Overview tab, click Share, share your layer with Everyone (public), and click OK.

16. Click the arrow next to Open in Map Viewer and click Add to new map with full editing control.

Open in Map Viewer ⌄
Add to new map
Add to new map with full editing control

The map viewer opens and the Change Style pane appears automatically. If the pane does not appear, you can click the Change Style button under the 311 Incidents layer.

17. In the Change Style pane, for the attribute to show, click Incident_Type. For drawing style, leave Types (Unique symbols) as the default, and click Options.

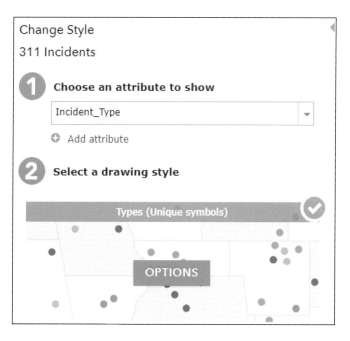

18. Click the symbol of each incident type, choose a symbol that users can intuitively understand according to the following list, and set its size to **24**.

- Dead Animal from the Outdoor Recreation set
- Graffiti from the Safety Health set
- Manhole Cover from the Cartographic set
- Pothole from the Transportation set
- Street Light from the People Places set
- Street Sign from the Transportation set

By using the unique value style, you have created a feature template. The existing incident types are the only values allowed.

19. For Visible Range, drag the slider handles to cover the full range.

20. Click OK and then Done to exit the Change Style panel.

The new symbols appear on the map.

21. In the Contents pane, point to the 311 Incidents layer, click the More Options button ⋯, and click Save Layer.

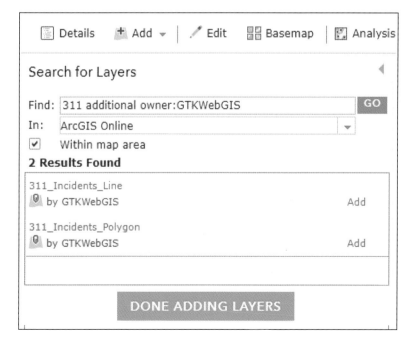

This step saves the symbols you just defined to the feature layer. In other words, these symbols will go with this layer. If you add this layer to another web map, you will not need to repeat the previous steps to define its style again.

You have a point layer in your web map. Next, you will add two pre-created feature layers, a line type and a polygon type, to your web map so you can practice collecting line and polygon features in the next section.

22. On the map viewer toolbar, click Add, select Search for layers.

23. In the Search for Layers pane, search for 311 additional owner:GTKWebGIS in ArcGIS Online, and the add links.

24. On the map viewer toolbar, click Save. Specify the Title, Tags, and Summary as illustrated.

Save Map

Title: 311_Incidents
Tags: 311 ✕   Incidents ✕   GTKWebGIS ✕
                Add tag(s)
Summary: For use with Getting to Know Web GIS book 3rd edition
Save in folder: webgis

SAVE MAP   CANCEL

25. Click Share ⊟, share your map with everyone, or certain groups that your field users belong to only, and click Done.

4.2 Collect data using Collector for ArcGIS

Collector for ArcGIS requires an ArcGIS Online or Portal for ArcGIS account, and isn't ideal for public users to use. You can create a browser-based app based on the web map you created previously, using the GeoForm or Edit web app template in ArcGIS Online and Portal for ArcGIS. If you share the web app with everyone, public users will be able to use this app to report 311 incidents. A sample web app using GeoForm is available at **http://arcg.is/1bmer5**. You can open it on your phone to report 311 incidents.

This section focuses on the Collector for ArcGIS mobile app. To install this app on your smartphone or tablet (iOS or Android), go to the App Store or Google Play store, search for **Collector for ArcGIS** and install it. This section follows the instructions for the iPhone version. You should find similar instructions for other phones but may find different tool layouts for tablet versions.

Note: At the time this book was written, there was a new version of Collector for ArcGIS being developed. This version will be the new Collector for ArcGIS. The version introduced in this section will likely be named Collector for ArcGIS Classic.

1. On your smartphone or tablet, tap the Collector app to start the app.

Collector

2. Tap ArcGIS Online or Portal for ArcGIS and sign in with your account. (If a prompt asks you to allow Collector for ArcGIS to access your location, click Allow.)

You can also enable location services for Collector in Settings > Privacy > Location Services > Collector > while using the app.

3. Find and tap the 311 Incidents web map you created in the last section.

4. Familiarize yourself with the functions of the buttons. Tap More to see the additional buttons.

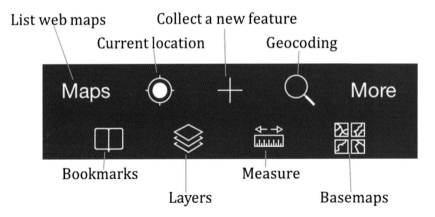

5. Tap the Collect a new feature button + to create a new incident. In the Collect a new feature panel, tap Pothole to choose this type of incident.

6. In the Attribute pane, notice that the incident Type is already set as Pothole, with a red exclamation warning indicating the symbol for potholes. Specify the rest of the attributes. For example, for Description, specify **Big pothole!**; for Date, choose today; for Your Name, specify your name; and for Your Phone, specify a phone number.

You can set the values as illustrated in the screen capture or make up your own.

7. Tap the Map button to see the location of this incident.

If you enabled location services for Collector for ArcGIS, your map should have opened with the extent reflecting your current location. The ability to locate your current location is convenient if you are at the place of the incident. Also, notice the icon with a number. This number indicates the margin of error at this moment. This margin of error is also indicated by the outer circle of the icon. If you are in an open space with GPS enabled on your mobile device, the margin of error may be only a few feet. If you are indoors, the margin of error can be much higher and the accuracy considerably lower.

Optionally, if you want to manually specify a location, you can tap the desired location on the map. To place the incident more precisely, you can tap and hold to get a magnifier. Use the magnifier to more precisely place the incident.

The result will be a red dot, which marks the location of the incident.

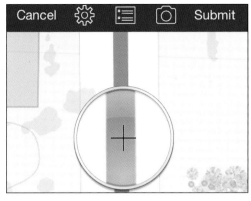

8.   On the toolbar, tap the Add Attachment button ⌾ to attach a photo or video to the incident.

9.   Tap Add > Take Photo or Video, and take a photo or video. Once you are happy with the picture, tap Use Photo; otherwise, retake the picture.

You can tap Add again if you want to attach more photos or videos to this incident.

10. Tap Done to finish adding attachments.

11. Tap Submit on the toolbar to save the incident.

You have collected a point, and the point is displayed with the correct symbol. Next, you will collect lines using both manual and stream modes.

12. While you are in the map mode, tap the Collect a new feature button to create a new incident. Choose a line feature type, such as the Street to Resurface incident type.

13. Specify attributes for this incident. For example, for Description, specify **Lots of cracks!**; for Date, choose today's date; for Your Name, specify a name; and for Your phone, specify a phone number.

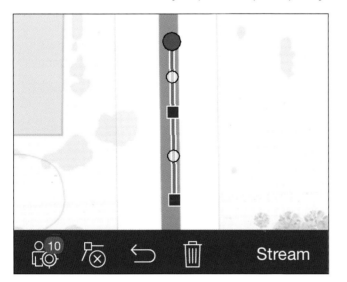

Cancel ⚙ 🗺 📷 Submit

／ **Length**
No valid Length

Incidents_Line

Type
Street to Resurface ＞

Description
Lots of cracks! ＞

Date
November 24, 2015 ＞

Your Name
Peter Smith ＞

Your Phone
123-4567890 ＞

14. Tap the Map button to switch to the map mode.

15. To add a line manually, tap the map to specify the nodes and vertices.

You can also collect lines or polygons by streaming your GPS location. First, you will review your streaming-related settings.

16. Tap the Settings button ⚙ to review the Collect Settings panel, including Required Accuracy and Streaming Interval. Change the values as needed.

If you are indoors, for example, the Required Accuracy often must be 20 meters, which is about 66 feet, or more, but if you are outdoor in open spaces, the Required Accuracy can often be set as 5 meters, which is about 16 feet, or less. The Streaming Interval can often be set as 5 seconds for walking.

- **Required Accuracy** sets the accuracy of data you must meet for the GPS to add a point.
- **Streaming Interval** sets how often a vertex is added to the feature you are creating. The smaller the time interval, the more detailed the shape.

Collect Settings	Done
Required Accuracy	20
Feet	Inches
Streaming Interval	5 sec

Choose the required location accuracy and interval for collection.

Smaller values improve the location accuracy and detail of the shape being collected.

17. Tap Done.

18. To add a line using GPS streaming, tap the Stream button, and take your phone while walking or driving along the line formation (such as a street or trail) where you wish to collect data.

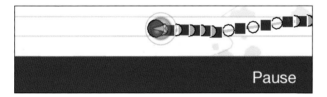

The GPS locations of your phone are collected and drawn on the map as you move.

19. In the same way that you did for points, tap the Add Attachment button to attach photos or videos relevant to the line incident.

20. Tap Submit on the toolbar to save the line incident.

You just collected a line feature. Similarly, you can collect a polygon feature using either the manual or stream mode.

You are now familiar with Collector for ArcGIS and can collect point, line, and polygon features. The collected data is saved to the feature layers that hold all the incident data. You can see the data you collected if you open the web map you created in section 4.1.

4.3 Design a survey for Survey123 for ArcGIS

In this section, you will design a survey form for initial disaster damage assessment.

1. Open a web browser, go to **http://survey123.arcgis.com**, and sign in.

2. Click Create a New Survey.

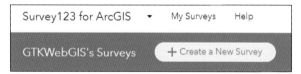

3. On the Using the web designer side, click the Get Started button.

4. In the Create a new Survey window, choose the following settings:
    - Click the thumbnail, browse to C:\EsriPress\GTKWebGIS\Chapter4, select Survey_Icon.png, and click Open.
    - For Name, specify **Initial Damage Assessment**.
    - For Tag, specify **GTKWebGIS, Damage Assessment**.
    - For Summary, specify **Collect initial property damage information**.

- Click Create.

The Survey123 website displays an empty survey, with a list of available question types on the right. For the first question, you will add a GeoPoint question for collecting property locations.

5.   Click the GeoPoint button to add it to the survey.

You will be directed to the edit tab of the Design pane.

6.   Under the Edit tab, choose the following settings:

- For Label, specify **Property Location**.
- For Default Map, choose Image with Labels.
- Select This is a required question.
- Click the Save button.

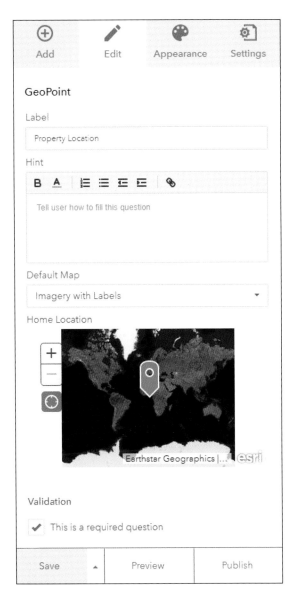

A GeoPoint question will use the device's location by default. This value can then be modified manually by the user.

7. In the Design pane, click the Add tab, and click Single Choice to add it to the survey.

8. Under the Edit tab, choose the following settings:

    - For Label, specify **Damage Severity**.
    - Set the choices as **Destroyed, Major, Minor, Affected,** and **Inaccessible**.
    - Select This is a required question.
    - Click the Save button.

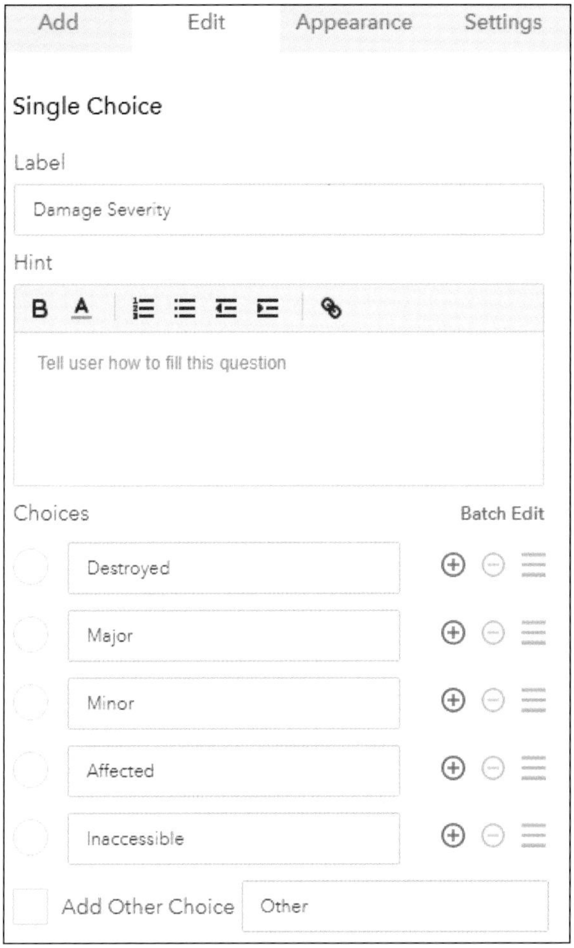

Add	Edit	Appearance	Settings

Single Choice

Label

Damage Severity

Hint

B **A**	≣ ≡ ≣ ≣	🔗

Tell user how to fill this question

Choices   Batch Edit

○ | Destroyed | ⊕ ⊖ ≡

○ | Major | ⊕ ⊖ ≡

○ | Minor | ⊕ ⊖ ≡

○ | Affected | ⊕ ⊖ ≡

○ | Inaccessible | ⊕ ⊖ ≡

☐ Add Other Choice | Other |

9. In the Design pane, click the Add tab, and click Multiple Choice to add it to the survey.

10. Under the Edit tab, choose the following settings:

   - For Label, specify **Damaged Elements**.
   - Set the choices as **Roof**, **Walls**, and **Foundation**.
   - Select Allow "Other."
   - Select This is a required question.
   - Click the Save button.

Next, you'll make the visibility of the damaged elements question conditional on whether the damage severity is Destroyed, Major, or Minor; other damage severity choices such as Inaccessible and Affected won't use this question.

11. Click the Damage Severity question and click the Set Rule icon that appears.

Damage Severity *

○ Destroyed

○ Major

○ Minor

○ Affected

○ Inaccessible

Set Rule

X   ▢   ⌐

Damaged Elements

☐ Roof

☐ Walls

☐ Foundation

☐ Other

12.   In the Set Rule window, choose the following settings:

- Under If, click the drop-down list to choose Destroyed, and under Show click the drop-down list to choose Damaged Elements.
- Under If, click the drop-down list to choose Major, and under Show click the drop-down list to choose Damaged Elements.
- Under If, click the drop-down list to choose Minor, and under Show click the drop-down list to choose Damaged Elements.
- Click OK.

Set Rule		✕
Show selected questions only if the answer to this question is a specific choice.		
If	**Show**	
Destroyed	Damaged Elements	⊖
Major	Damaged Elements	⊖
Minor	Damaged Elements	⊖
-Please Select- ▼	-Please Select- ▼	

Next, you will add an Image question to allow users to upload a photo of the damaged property.

13.   In the Design pane, click the Add tab, and click Image to add it to the survey.

14.   Under the Edit tab, choose the following settings:

- For Label, specify **Photo of Damages**.
- Select This is a required question.
- Click the Save button.

15.   In the Preview pane, drag and drop the photo question to the space under Damaged Elements question.

Next, you will add a question to ask for the assessor's name.

16. In the Design pane, click the Add tab, and click Singleline Text to add it to the survey.

17. Under the Edit tab, choose the following settings:

    - For Label, specify **Assessor name**.
    - Select This is a required question.
    - Select Cache answer to this question.
    - Click the Save button.

Next, you will add a question to ask for the assessor's email.

18. In the Design pane, click the Add tab, and click Email to add it to the survey.

19. Under the Edit tab, choose the following settings:

    - For Label, specify **Assessor email**.
    - Select This is a required question.
    - Select Cache answer to this question.
    - Click the Save button.

Next, you will add a question to record the assessment date.

20. In the Design pane, click the Add tab, and click Date to add it to the survey.

21. Under the Edit tab, choose the following settings:

    - For Label, specify **Assessment date**.
    - For Default Value, Select Submitting date.
    - Select This is a required question.
    - Click the Save button.

Optionally, you can click the Preview button to preview your survey to see how it will look on desktops, phones, and tablets.

22. Click Publish. In the Publish Survey window click Publish again.

23. If you see a window saying your survey has been published, click OK to continue.

Next, you will share your survey form with your audience.

24.   Click the Collaborate tab.

25.   Share your survey with everyone, your organization, or only certain groups, and click the Save button.

26.   For Survey link, click Ask the user how to open the survey, in the browser or in the Survey123 field app.

Your survey form is ready. You can email your audience the URL and the QR code (click the QR Code button ▦ to bring up the QR code), or post the link and code to your social media websites or your organization's website.

27.   Keep this page open for use in the next section.

4.4 Use Survey123 to collect data and review the data collected

Survey123 has a browser-app version and a native-app version. You can use either version on computers, tablets, and smartphones. This section will use the iPhone native edition. Other editions are similar to the native edition.

1.   If you haven't installed Survey123, search for Survey123 in the App Store on your iPhone or Google Play on your Android phone, and install it.

2.   Continuing from the last section, click the QR Code button ▦ to show the QR code.

Scan the QR code to open the survey
on your device

Survey link:

https://survey123.arcgis.com/share/8f49ef39a86245918cd7302b3ab0b80e?open=menu

Open the survey in browser directly

● Ask the user how to open the survey, in browser or in the Survey123 field app

Open the survey in the Survey123 field app directly.   (Learn more about this option)

If you have left the page, you can go back to the page by signing in at **http://survey123.arcgis .com**, clicking My Surveys, finding the survey you created in the previous section, and clicking the Collaborate button .

3. On your phone, start the Camera app, and point your phone to the QR code.

On recent iOS, the Camera app is also a QR scanner. If you don't have a QR scanner on your phone, you can find a QR scanner app in the App Store or Google Play store, install the app and use it to scan the QR code for your survey.

4. To open the survey in the Survey123 field app, follow the instructions on your phone.

Your survey is loaded in the app. Next, you will use the survey to collect data.

5. On the form, specify the following information:

• For Property Location, tap My Location button to use your current location.

You can also tap, zoom, and pan the map to place the pin at a location different from your current location, and go back to the form.

• For Damage Severity, select Destroyed, Major, or Minor. Notice the Damaged Elements question appears.

- For Damaged Elements, choose one or more options.
- For Photo of Damages, tap the Camera icon to take a photo or tap the Folder button to select a photo on your phone.
- For Assessor name, type a name.
- For Assessor email, type an email address.
- For Assessment date, leave the date as the default (the current date).
- Tap the Submit button (the green check mark), and tap Send Now.

Next, you will collect one more record.

6. In the Survey123 mobile app, under My Surveys, tap Initial Damage Assessment.

7. Notice you can collect new data and review the data you have collected.

8. Tap Collect.

9. Repeat step 5 to collect a new record. For Damage Severity, choose Affected or Inaccessible, and notice the Damaged Elements question doesn't show up. Submit the record now.

Next, you will review the data you collect on the Survey123 home page.

10. On your computer, open a web browser, go to **http://survey123.arcgis.com**, sign in, click My Surveys, and find and click the survey you were using to open it.

11. On the Overview page, notice the total records collected, the total participants, and the time the records were collected.

12. Click the Analyze tab, and review the charts and summaries on the answers to Damage Severity, Damaged Elements, and Assessor names.

13. Click the Data tab, review the locations of the records on the map, and the attributes in the table. Click a record on the map or in the table, and notice the individual responses appear.

14. Click the Export drop-down menu and select CSV as the data download format.

The survey data should automatically begin to download.

15. Open the data downloaded in Excel or a text editor to see the data collected.

4.5 Create native apps using AppStudio for ArcGIS

In this section, you will create a native app that allow users to report non-emergency incidents.

1. Open a web browser, go to **http://appstudio.arcgis.com**, and sign in.

2. Click My Apps.

3.   Click Create New App.

You will see several available templates. The number of templates will grow with future releases.

4.   Find the Quick Report template and click Start with this template.

5.   On the App Info page, set the Title as **Incidents Reporter**.

Optionally, you can change the Thumbnail, Launch Image, and App Icon.

6.   Click Save to save your settings.

7.   Click Quick Report Settings.

8.   Click Choose Feature Service.

9.   In the Choose Feature Service window, click the Public tab.

10. Find the 311 Incidents feature layer you created in section 4.1, or Search for **311 incidents 3rd edition owner:GTKWebGIS** under the Public tab.

11. Click to select the resulting item, and then click Next.

12. Click the 311 Incidents feature service to expand it, click 311 Incidents (point) layer, and then click OK.

13. Click Save.

Next, you will preview your configured app.

14. Click View the App Live and read the instructions in the pop-up.

  • On your mobile device, go to Google Play or the App Store, search for **AppStudio Player for ArcGIS**, and download and install the app.
  • On your mobile device, start AppStudio Player for ArcGIS, and sign in with the same ArcGIS Online account you used at beginning of this section.
  • From the app list, find and select the app you just created to download it, open it, and preview your app.

15. In the AppStudio web browser, click OK to close the View the App Live window.

16. Click Save and Finish, which brings up the App Console page.

17. Click Build App.

18.   Choose the platform(s) you want for building your app. If you do not have a mobile device, click a desktop platform.

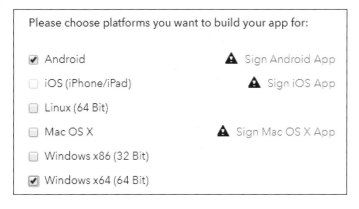

19.   Click Build.

Your request to build the app enters a queue for processing. Building your app may take a few minutes.

4.6 Deploy and test your native app

Building your app will create an installation package for the platform you selected. After your build is complete, you will need to install the installation package on your device. It's relatively easy to deploy your app on an Android device. If you don't have an Android device, you can deploy the app using a computer version, which works similarly to the mobile native versions.

1.   If you are using a computer, skip to step 7.

2.   Complete the following actions if you choose to use an Android device:

- Continue from the previous section, and click the QR code icon to get the QR code.
- Use your camera or another QR Reader to scan the QR code.
- Follow the instructions to download IncidentsReporter.apk.

3. Tap Apps, tap Downloads, find the installation file you just downloaded, and tap the file to run it.

If you did not enable Unknown sources for your Android to Install Apps Outside the Play Store, you will be prompted with a message saying Install Blocked.

4. Click Settings, and in the Unknown sources window, select Allow initial installation only, and click OK.

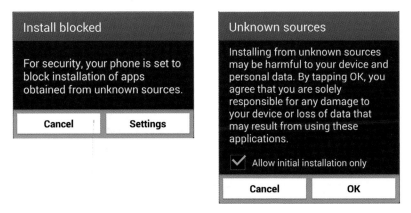

5. Follow the instructions to install the app.

6.   On your Android device or desktop computer, find the Incidents Reporter app you just installed, start the app, and skip to step 8.

7.   On your computer, continue from the last section, click the Download button for your computer OS, find the downloaded file, extract the file, launch the setup file, (Ignore warnings saying that this file is an unrecognized app or is from an unknown publisher), finish the installation, and run the app.

You will first see the initial splash image along with two buttons.

8.   Tap or click the New button.

9.   Pick an incident type, and then tap Next.

10. Add an incident location by using your current location or manually selecting a location on the map and tap Next.

11. Add photo(s) using your camera or from your gallery. Tap Next.

12. Fill out the incident attribute form.

13. Tap Submit.

Once you are satisfied with the app you created using AppStudio for ArcGIS, you will typically authorize your app, and submit the app to Google Play and the App Store for your users to download. You also can release the app for enterprise deployment without going through Google Play and the App Store. Refer to the AppStudio for ArcGIS website for details about how to deploy your apps.

In this tutorial, you first configured an editable feature layer, and used Collector for ArcGIS to collect points, lines, and polygons data. You then created a smart survey using Survey123 web designer and used it with Survey123 for ArcGIS. Then, you created a Quick Reporter native app using AppStudio for ArcGIS, deployed it, and collected data using your native app.

QUESTIONS AND ANSWERS

1. What are the differences among Collector for ArcGIS native app, Survey123 for ArcGIS native app, and ArcGIS GeoForm web app template?

   Answer: See the table. Refer to **http://arcg.is/2ySDVwA** (short for **https://community.esri.com/groups/survey123/blog/2015/09/04/survey123-collector-and-geoform-a-quick-comparison**) for more details.

	Survey123 for ArcGIS	Collector for ArcGIS	GeoForm
Data collection style	Form centric	Map centric	Form centric
Supports capturing new data	Yes	Yes	Yes
Supports editing existing data	Yes	Yes	No
Smart forms	Yes (xForms)	No (the classic edition) Yes (the new edition)	No
Works offline	Yes	Yes	Yes*
Supports anonymous access	Yes	No	Yes
Platforms	iOS, Android, Windows (7,8,10), Mac, Linux, Web	iOS, Android, Windows 10	Web

*Geoform supports offline capabilities if your web browser stays open.

2. Do I need to be an ArcGIS named user to use the Survey123 field app or web form?

   Answer: Not for public surveys. Once a survey is made public, anyone who has the link to the survey can submit data to it. No ArcGIS account is required.

3. I created an editable feature layer and added it to a web map. I shared both the web map and the layer with everyone. But my colleague still can't find the web map in her Collector for ArcGIS. Why?

   Answer: You must share the web map with a group to which your colleague belongs.

4. Can I work with related records in Collector?

   Answer: Yes. Collector for ArcGIS supports editing related features and tables. When using related data, it is strongly recommended that you use GlobalIDs when defining relationships. User-maintained relationships are not recommended.

5. Can my users open Collector from my emails, websites, or apps?

   Answer: Yes. You can build a URL that starts with the identifier "arcgis-collector" and append additional parameters. The following are several examples. Refer to **http://bit.ly/2xr7f9U** (short for **https://github.com/Esri/collector-integration**) for more details.
   - The following URL opens Collector with a web map: **arcgis-collector://?item ID=35b1ccecf226485ea7d593f100996b49**
   - The following URL opens Collector with a web map and centers the map to a location: **arcgis-collector://?itemID=35b1ccecf226485ea7d593f100996b49& center=34.0547155,-117.1961714**
   - The following example URL initiates a feature collection for a layer, and the center is used to define the new feature's geometry: **arcgis-collector ://?itemID=5d417865c4c947d19a26a13c7d320323&center=43.524080, 5.445545&featureSourceURL=http://sampleserver5a.arcgisonline.com/ arcgis/rest/services/LocalGovernment/Recreation/FeatureServer/0**

ASSIGNMENT

Assignment 4: Use Survey123 to engage citizens to report wanted suspects.

Law enforcement typically invites citizens to report wanted suspects through web pages or phone calls. The police department in your city wants to enhance the workflow. The police would like to provide a mobile app and a web app for citizens to report where they have seen four highly dangerous suspects and upload photos and videos that include images of these suspects.

Requirements: Your survey should have the following fields:
- Suspect name: Allow users to choose from a list.
- GeoPoint: Location where you saw the suspect.

- Description: Describe what the suspect was doing and where you saw the suspect.
- Date: When you saw the suspect.
- Image.
- Your name: Optional. Users can choose to be anonymous.
- Your contact info.
- Report date.

What to submit: The URL that asks your users to open the survey in a browser or in the Survey123 field app.

Resources

"5 Minutes to Your First Collector Map," https://community.esri.com/community/gis/applications/collector-for-arcgis/blog/2017/10/03/5-minutes-to-your-first-collector-map (or http://arcg.is/2xZSoG2).

"5 Things to Try in Collector's Aurora Beta," https://community.esri.com/community/gis/applications/collector-for-arcgis/blog/2017/12/05/5-things-to-try-in-collector-s-aurora-beta (or http://arcg.is/2nv8zrG).

"APIs, SDKs and Apps," https://developers.arcgis.com/documentation/core-concepts/apis-sdks-apps.

"AppStudio for ArcGIS," https://appstudio.arcgis.com.

"AppStudio for ArcGIS: An Introduction," Chris LeSueur, Nakul Manocha, Anshuman Mathur, and Loayeh Jumbam, https://www.youtube.com/watch?v=QFrzNbIuIFA&list=PLaPDDLTCmy4YcXpv_ypX3YicMHVUOuGYR (or http://bit.ly/2y8ekQU).

"Chapter 2, The Power of Apps," *The ArcGIS Book*, Christian Harder and Clint Brown, http://learn.arcgis.com/en/arcgis-book/chapter7.

"Collector for ArcGIS," http://doc.arcgis.com/en/collector.

"Collector for ArcGIS: An Introduction," https://www.youtube.com/watch?v=PuMbhhoPqrU&index=33&list=PLaPDDLTCmy4YcXpv_ypX3YicMHVUOuGYR (or http://bit.ly/2zbWQTR).

"Collector—The Aurora Project," https://community.esri.com/community/gis/applications/collector-for-arcgis/blog/2017/09/06/the-aurora-project (or http://arcg.is/2fx1q3b).

"Get Started with Survey123 for ArcGIS," http://learn.arcgis.com/en/projects/get-started-with-survey123.

"Manage a Mobile Workforce," http://learn.arcgis.com/en/projects/manage-a-mobile-workforce.

"Manage and Complete Hydrant Inspections," http://learn.arcgis.com/en/projects/manage-and-complete-hydrant-inspections.

"Optimizing Field Operations with ArcGIS Apps," https://www.youtube.com/watch?v=WLDPi_dawxs&list=PLaPDDLTCmy4YcXpv_ypX3YicMHVUOuGYR (or http://bit.ly/2gH7rek).

"Smart Sketching in Survey123: Stroke by stroke," Ismael Chivite, https://community.esri.com/groups/
    survey123/blog/2017/10/09/smart-skething-in-survey123-stroke-by-stroke (or http://arcg.is/
    2wOxDKx).

"Survey123 for ArcGIS," http://survey123.esri.com.

"Survey123 for ArcGIS: An Introduction," James Tedrick, Marika Vertzonis, and Zhifang Wang,
    https://www.youtube.com/watch?v=YJ8HdIqBG7w&index=34&list=PLaPDDLTCmy4YcXpv_
    ypX3YicMHVUOuGYR (or http://bit.ly/2lm10lR).

CHAPTER 5

Tile layers, map image layers, and on-premises Web GIS

In previous chapters, you learned how to create hosted layers and web apps using ArcGIS Online, which runs on a public cloud. There are situations when you need to build on-premises (locally hosted) or on-premises/cloud hybrid Web GIS. The reason may be security or functionality. You can build on-premises and hybrid Web GIS using ArcGIS Enterprise. This chapter will compare the similarities and differences between ArcGIS Online and ArcGIS Enterprise. This chapter then introduces three basic types of web layers: vector tile, raster tile, and map image; explains the scenarios in which each fits; and compares them with feature layers. The tutorial teaches how to connect ArcGIS Pro to portals; how to publish vector tile, raster tile, and map image layers; and how to create a side-by-side comparison web app to compare the performances of vector tile, raster tile, and map image layers. ArcGIS Online and core ArcGIS Enterprise share more similarities than differences. Most of the content in this chapter applies to both online and on-premises Web GIS.

Learning objectives

- *Understand the need for on-premises and hybrid Web GIS.*
- *Know ArcGIS Enterprise and its components.*
- *Understand the concept of portal collaboration.*
- *Know the differences and similarities of vector tile, raster tile, and map image layers.*
- *Become familiar with the workflow to publish web layers using ArcGIS Pro.*
- *Explore and use web layers in web maps.*
- *Create web apps for side-by-side comparison.*

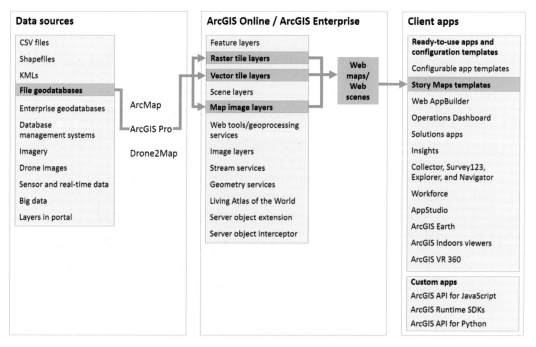

ArcGIS offers many ways to build web apps. The green line in the figure highlights the technology presented in this chapter.

The need for on-premises and hybrid Web GIS

In previous chapters you have learned about ArcGIS Online as an easy and excellent entry point to Web GIS. ArcGIS Online runs in an Esri-managed cloud infrastructure in an SaaS model. The computing and data storage resources, the website, and every other aspect of the system is hosted by Esri in a highly scalable environment. Organizations use ArcGIS Online to create, use, and share maps and apps internally and with the public.

However, organizations sometimes must have ArcGIS Enterprise. Core ArcGIS Enterprise and ArcGIS Online provide similar capabilities, but Core ArcGIS Enterprise runs on organizations' internal infrastructures. These infrastructures can be owned and operated directly by these organizations or, optionally, managed by other cloud providers, such as Amazon Web Services or Microsoft Azure. Organizations typically implement ArcGIS Enterprise for the following situations:

- **The need for on-premises Web GIS:** Some organizations may not have Internet connections, or their connection may be prohibited or unreliable. Government or corporate regulations can prevent organizations from using public cloud services or storage. These situations require on-premises Web GIS.
- **The need for hybrid Web GIS:** Many organizations would like to keep their own operational layers on their own infrastructure while also using ArcGIS Online content, such as the basemaps, Living Atlas of the World layers, GeoEnrichment, and other analysis services. Other organizations would like to have their own operational layers referenced in ArcGIS Online web maps and web apps. These situations require hybrid Web GIS.
- **The need for functionalities only available with ArcGIS Enterprise:** ArcGIS Online provides a collection of geoprocessing services, image services, stream services, and geocoding services. Due to considerations of security and performance, ArcGIS Online doesn't yet allow organizations to publish such types of services of their own. Organizations now must have ArcGIS Enterprise to publish such types of services.

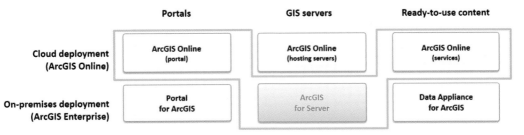

Web GIS deployment patterns: The cloud deployment row represents all portal, GIS servers, and ready-to-use contents in the public cloud. The on-premises deployment row represents all components that are within an organization's internal infrastructure. The shaded combination represents one form of a hybrid deployment pattern.

ArcGIS Enterprise and organizational subscriptions to ArcGIS Online are complementary implementations of Web GIS. They provide similar functionalities, such as a portal website with map and scene viewers, a content repository, and the same API for developers. They also support similar types of web services or layers, with some differences.

ArcGIS Online and ArcGIS Enterprise share similar functionalities with some differences in their support of various web layers or services.

	Feature layer	Raster tile layer	Vector tile layer	Map image layer	Scene layer—multipatch and points	Scene layer—point cloud	Imagery layer	Web tool	Stream services
ArcGIS Online	✓	✓	✓	✗	✓	✓		○	○
ArcGIS Enterprise	✓	✓	✓	✓	✓	✓	✓	✓	✓

✓ = This layer type is supported through regular sharing workflows.

✓ = This layer type requires a package to be created manually before sharing.

○ = ArcGIS Online provides rich collections of services or layers of this type, but users can't publish services or layers of this type.

✗ = This layer type is not yet supported.

ArcGIS Enterprise components and deployment

ArcGIS Enterprise is Esri's product for on-premises and hybrid Web GIS. You can think of ArcGIS Enterprise as an ArcGIS Online, with some differences, that can be deployed to your servers, your private cloud, or the public cloud. ArcGIS Enterprise has a flexible deployment model that supports various configurations:

- On-premises, on physical hardware or virtualized environments
- In the cloud, on Amazon Web Services (AWS) and Microsoft Azure
- A mixture of on-premises and cloud

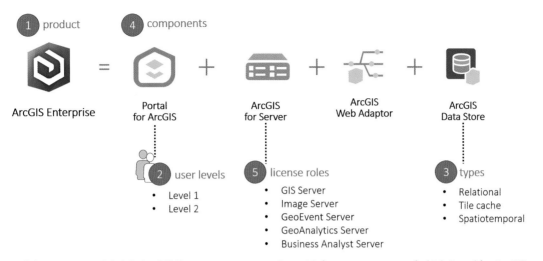

ArcGIS Enterprise 1-2-3-4-5: ArcGIS Enterprise is one product with four components, of which Portal for ArcGIS has two user levels. ArcGIS Data Store has three types, and ArcGIS Server has five license roles.

ArcGIS Enterprise includes the following software components that are designed to work together:

- **ArcGIS Server:** ArcGIS Server can create and host various types of geospatial web services, which allow a powerful server computer to receive and process requests sent by various clients. ArcGIS Server is the core component that makes your geographic information available to others in your organization and, optionally, anyone with an Internet connection. ArcGIS Server has the following five license roles:
  - **ArcGIS® GIS Server:** Powers traditional GIS web services and layers.
  - **ArcGIS® Image Server:** Supports image services and online raster analysis. Refer to the image services chapter for more details.
  - **ArcGIS® GeoEvent Server:** Powers real-time GIS including tracking things, creating geofences, and performing analysis in real time. GeoEvent Server can produce outputs in feature layer and stream services. Refer to the real-time GIS chapter for more details.
  - **ArcGIS® GeoAnalytics Server:** Performs spatial temporal analyses on small to large datasets. Refer to the online analysis chapter for more details.
  - **ArcGIS® Business Analyst Server:** Provides advanced capabilities for business analysis.

- **Portal for ArcGIS:** Portal for ArcGIS is essentially a geospatial content management system. It brings together all the geographic content, including layers, maps, and apps, in your ArcGIS platform and shares it throughout your organization. Like ArcGIS Online, Portal for ArcGIS users have levels, roles, and privileges. Level 1 is for members who only need privileges to view content and join groups. Level 2 is for members who need to view, create, and share content and own groups, in addition to other tasks. Portal for ArcGIS provides the following capabilities:
  - Create, save, and share web maps and scenes.
  - Create and host web mapping apps.
  - Search for GIS content within your organization.
  - Secure the access to your GIS content.
  - Manage organizational utility services.
- **ArcGIS Data Store:** This is an application that lets you easily configure data storage for hosting and federated servers used with your portal. ArcGIS Data Store has the following different types of data stores:
  - **Relational data store:** stores data for your portal's hosted feature layers.
  - **Tile cache data store:** Stores caches for your portal's hosted scene layers.
  - **Spatiotemporal big data store:** Archives real-time observational data that you can use with an ArcGIS Server running ArcGIS GeoEvent Server that is federated with your portal. It also stores the results generated using ArcGIS GeoAnalytics Server tools.
- **ArcGIS Web Adaptor:** ArcGIS Web Adaptor is an application that can be deployed to your existing website and forward requests to your ArcGIS Enterprise machines. This web adaptor allows you to integrate your ArcGIS Server and Portal for ArcGIS with your existing web server and your organization's security mechanisms.

Base ArcGIS Enterprise deployment

Base ArcGIS Enterprise deployment is a foundational setup of ArcGIS Enterprise with all four components installed and configured in a certain way for them to work together. The base deployment consists of the following requirements:

- ArcGIS Server must be licensed as ArcGIS GIS Server Standard or ArcGIS GIS Server Advanced and configured as the hosting server for your portal.
- Portal for ArcGIS must be federated with ArcGIS Server. As the result, Portal for ArcGIS manages the security and access to ArcGIS Server.
- ArcGIS Data Store should be configured as a relational and tile cache data store.
- One ArcGIS Web Adaptor installation for traffic to Portal for ArcGIS and another one for traffic to ArcGIS Server.

Base ArcGIS Enterprise deployment deploys all four ArcGIS Enterprise components configured to work together. Portal for ArcGIS and ArcGIS Server are federated. ArcGIS Server is configured as the hosting server for Portal for ArcGIS.

Base ArcGIS Enterprise deployment supports the following flexible deployment scenarios:
- **Single-machine deployment:** All components are installed on a single machine or virtual machine.
- **Multitiered deployment:** Each component is installed on a separate physical or virtual machine.
- **Highly available deployment:** Each component is configured with redundancy to minimize downtime in scenarios where one or more machines become unavailable.

Additional license roles require the base deployment and can be added to the base deployment. ArcGIS Enterprise is also scalable. The hosting server site in ArcGIS Enterprise deployment can be scaled out with additional server resources as your user base and their use of analytical features increases. ArcGIS Enterprise provides the following tools to streamline and automate the deployment process:

- ArcGIS Enterprise Builder
- ArcGIS Chef cookbooks
- ArcGIS® Enterprise Cloud Builder on Amazon Web Services
- ArcGIS® Enterprise Cloud Builder for Microsoft Azure

Portal collaboration

Organizations are naturally distributed, and the structure of organizations is often hierarchical. Effective collaboration between these distributed organizations is key to their success. For example, if a city has many bureaus, and each bureau has many departments, then every bureau and department can have its own ArcGIS Enterprise or ArcGIS Online. To achieve their common goals/initiatives, these organizations must share their curated content with each other. With ArcGIS Online and ArcGIS Enterprise, sharing content within organizations is typically accomplished using groups. Sharing content with external Web GIS deployments can be accomplished using distributed collaboration.

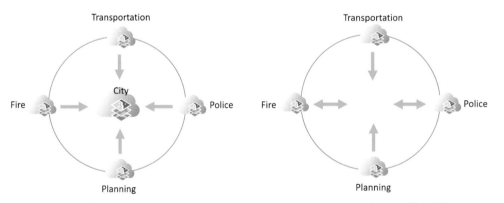

Two scenarios of distributed collaboration. Each participant organization has its own Web GIS deployment. The organizations can be configured to send, or receive, or send and receive GIS content.

The benefit of establishing distributed collaboration is to organize, network, and share content across organizations. Distributed collaboration is based on a foundation of trust and driven by common goals or initiatives. Once a trusted collaboration among collaboration participants is established, these participant organizations can replicate or synchronize content across the network of participants. The synchronization can happen immediately as the content in the source is changed, or can happen at scheduled intervals, for instance, every 24 hours. The content types that can be replicated include layers, web maps, and web apps. The replicated content becomes discoverable inside each organization participating in the collaboration.

Distributed collaboration provides a secured solution for different Web GIS deployments, ArcGIS Online, or ArcGIS Enterprise, to share contents. Distributed collaboration creates distributed Web GIS, a new pattern of a GIS deployment.

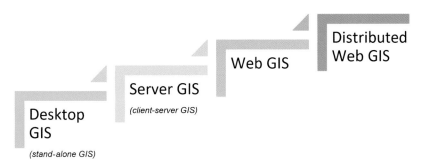

GIS deployment patterns have evolved from desktop GIS, to server GIS, to Web GIS, and to distributed Web GIS. Web GIS of different departments and organizations can share contents securely and collaborate automatically.

Web layers and web services

ArcGIS Enterprise can create and host a suite of web layer types, including feature, vector tile, raster tile, map image, imagery, and scene. These layers are items in Portal for ArcGIS, and reference various types of web services, as illustrated in the next table.

Web layers and their corresponding web services

Web layer	Web service
Map image layer	Dynamic/cached map service
Feature layer	Feature service
Tile layer	Cached map service (hosted)
Vector tile layer	Vector tile service
Scene layer	Cached scene service
Imagery layer	Image services

Feature layers have been introduced in previous chapters, while scene and imagery layers will be introduced in later chapters.

Raster tile layer

Raster tile layers, or tile layers, deliver maps to client applications as image files (for example, JPG or PNG format) that have been pre-rendered and stored on the server and are displayed in those formats by the client. Raster tile layers are most appropriate for basemaps or maps with content that is relatively static and hardly changes over time. Raster tile layers can take some time to create in advance and can require a huge amount of disk space to store. However, once created, tile layers can significantly improve the performance, availability, scalability, and the user experience of your web apps. Raster tile layers include the following benefits:

- Work well across a wide range of applications and devices (web, desktop, and mobile), including older versions of web browsers
- Provide high-end cartographic capabilities such as advanced label placement and symbology
- Support various raster data sources such as imagery and elevation data
- Can be printed from web mapping applications

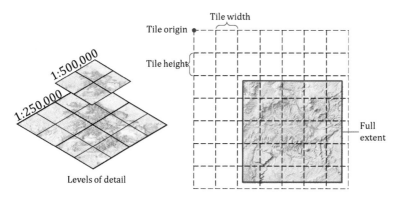

Raster tile layers generate a set of map tile images in advance at predetermined scale levels. The map caching scheme (right) includes the number of scale levels, the scale at each level, tile dimension, tile origin, tiling area, and image format.

Vector tile layer

Vector tile layers deliver map data as many vector files, usually in Protocol-Buffer Binary Format (PBF). These files are rendered on the client application based on a style delivered with the data. Vector tile layers include the following advantages:

- Map style can be customized. For example, users can hide the layer visibility, change symbols and fonts, change languages for labels, and so on, without having to regenerate tiles.
- Labels in vector tile can remain upward while the map rotates. This capability fits the needs of many scenarios.

- Vector tile layers can adapt to the resolution of the display device and look sharper on high-resolution displays (for example, retina devices) than low-resolution (typically 96 dpi) raster tiles.
- The size of vector data is usually smaller than raster data. In addition, vector tile layers typically generate tiles based on data density. There is no need to divide a tile into further levels if the tile encloses no more than the specified number of vertices. As a result, vector tiles are much smaller in size than raster tiles, and can be generated much more quickly, and with fewer hardware resources, than corresponding raster tiles. Generating smaller files reduces the cost to generate the tiles and improves the speed at which data updates can be made available.

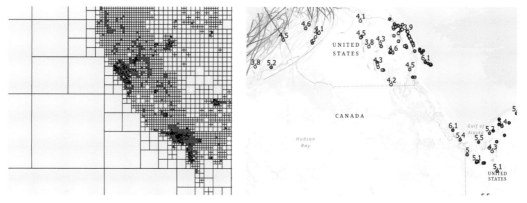

Left: Vector tile layers typically generate tiles based on data density. There is no need to divide a tile into further levels if the tile encloses no more than the specified number of vertices. Right: Labels in vector tiles can remain upward while the map rotates.

Map image layer

Map image layers can be drawn dynamically by the server or using tiles from a cache. Even when map image layers are using tiles, the vector data behind is still available, and thus map image layers support both visualization and spatial/attribute query. Map image layers and feature layers are appropriate for visualizing operational layers.

Feature, vector tile, raster tile, and map image layers are commonly used for displaying maps. They complement each other and fit for different uses. You should consider the following choices when selecting a layer type for your data:

- If the layer will be used for editing workflows, select feature layer.
- If the data changes regularly and will not be used for editing workflows, select feature layer or map image layer.
- If the data is static and will only be used for visualization, select raster tile layer, or vector tile if your data is vector type only.

- If your data is vector and will be used for operational layers, consider feature layer first. By default, feature layer is rendered using Scalable Vector Graphics (SVG), which limits the number of features that can be displayed in the view. ArcGIS Online and ArcGIS Enterprise are rolling out the support of WebGL-rendered feature layers, which has overcome the limitation of SVG technology. WebGL-rendered feature layers allow users to display hundreds of thousands, potentially millions, of features in web browsers quickly.

A comparison of basic web layers

	Feature layers	Map image layers		Raster tile layers	Vector tile layers
		Dynamic	Tiles		
Typically used for	Operational layers			Basemaps	
Editable	Yes (editing can be enabled or disabled)	No			
Fit for data sources	Dynamic and static data		Relatively static data		
	Vector data only	Raster and vector data			Vector only
Maps rendered on	Client side	Server side			Client side
Performance	Typically fast, WebGL feature layers can display millions of features in the browser	Slower	Faster		
Query support	Yes	Yes		No	
Pop-up support	Yes, faster (attributes are already on client side; no need to query server)	Yes, slower (must query server to retrieve attributes)		No	

Interoperability and web service standards

GIS web services have standards. Standards specify the interface that different vendors should use and are an important way to achieve interoperability among different vendors. Vendors can develop products against the standards independently, without knowing each other, and their resultant products will still be interoperable. It's the same principle as power plugs and outlets, which conform to the same specification of shape and voltage, so that you can simply "plug and play." Web service standards specify the format of HTTP requests and HTTP responses, including what parameters should be in a request, the name of each parameter, what type of value(s) is expected for each parameter, and what type of results should be in the response.

In the geospatial world, Open Geospatial Consortium (OGC) is the main standards body. OGC has defined a set of GIS web services standards including WMS, WMTS, WFS, Web Coverage Service (WCS), and Web Processing Service (WPS). OGC Keyhole Markup Language (KML) and GeoRSS (Really Simple Syndication) are not exactly web services standards, but they are often used in GIS web services to transport information between web services and their clients. ArcGIS web services comply with these OGC web services standards and thus are open.

Main types of GIS web services and their corresponding OGC standards that are supported in ArcGIS

Name in ArcGIS Server/ Online and Portal	Capabilities	ArcGIS Online	ArcGIS Enterprise	OGC Standards
Feature service/layer	Allows clients to retrieve vector features from the server and update the features on the server.	✓	✓	WFS
Tile map service/raster tile layer	Serves map image tiles, such as JPG, PNG, or GIF, which are generated in advance.	✓	✓	WMTS
Vector tile service/layer Serves pre-generated tiles of map in vector format.		✓	✓	
Dynamic map service/ map image layer	Generates and serves map images, such as JPG, PNG, or GIF, upon client requests.	✗	✓	WMS, KML
Scene service/layer	Serves 3D maps.	✓	✓	
Image service/layer	Provides access to the contents of a raster dataset or mosaic dataset, including pixel values, properties, metadata, and bands. ArcGIS image service additionally supports rapid map algebra calculation.	O	✓	WCS, WMS, WMTS
Geocode service/locator	Converts addresses and place names into X, Y locations.	O	✓	
Geoprocessing service/ web tool	Shares server's workflow and analysis functions with clients.	O	✓	WPS
Stream service	Pushes data to clients.	O	✓	
Geometry service	Supports geometric calculations such as buffering, simplifying, calculating areas and lengths, and projecting.	O	✓	

📓 **Notes:**

O indicates that this product provides services of this type but doesn't allow users to publish their own services of this type.

✓ indicates that this product allows users to publish their own services of this type.

✗ indicates that this product doesn't provide and doesn't allow users to publish services of this type.

Workflow to share web layers from ArcGIS Pro

You can publish web layers from ArcGIS Desktop, which include ArcMap and ArcGIS Pro. ArcGIS Pro provides tools to visualize, analyze, compile, and share GIS data in both 2D and 3D environments. ArcGIS Pro has better integration with ArcGIS Enterprise and can publish more types of web layers than ArcMap. For these reasons, this book uses ArcGIS Pro for publishing web layers. This chapter focuses on tile layers, vector tile layers, and map image layers. Refer to other chapters for scene layers and image layers.

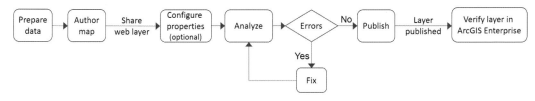

Workflow to create web layers using ArcGIS Pro.

Using ArcGIS Pro to create web layers involves the following three steps:
1. Prepare your data using ArcGIS Pro. The following practices are recommended to increase map service performance:
    *   For most situations, project your data into Web Mercator so your map aligns with the popular basemaps without having to be projected dynamically.
    *   Create indexes on the attribute fields that your users will query in your app.
    *   Turn off or delete attribute fields that are irrelevant to your app objectives.
2. Author the map document in ArcGIS for Desktop. Add data layers to your map. Configure layer symbols and other properties.
    *   Remove unused layers, including basemap layers.
    *   Do not use complex symbols that are not supported or will slow down your web layer performance.
3. Share web layer.
    *   Select copying data to server or referencing registered data sources. For ArcGIS Online, your data will always be copied to the server.
    *   Select an appropriate layer type.
    *   If needed, configure such properties as a tiling scheme.
    *   Analyze your map and review for any errors, warnings, and messages that may result. Once the errors are fixed, you can publish your web layer.
    *   Verify your web layers. After you publish the layer, go to Portal for ArcGIS, and add your layer to a web map to verify whether the layer works correctly.

ArcGIS Pro needs connections to a portal, either ArcGIS Online or Portal for ArcGIS. You must have an active portal connection. Typically, when sharing web layers from ArcGIS Pro, the data is either copied to the server side or referenced by the server side. The service is created in ArcGIS Server and registered in Portal for ArcGIS, which results in a web layer item in Portal for ArcGIS.

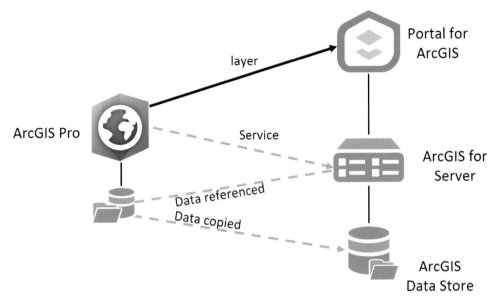

When you share web layers from ArcGIS Pro to Portal for ArcGIS, the data is copied to the server or referenced by the server. The service is typically created in ArcGIS Server and registered in Portal for ArcGIS, and results in a web layer item in Portal for ArcGIS. A similar process happens when sharing web layers to ArcGIS Online.

Make your data accessible to ArcGIS Enterprise

ArcGIS Enterprise needs access to your data so it can serve your web layers, for example, generate maps and respond to queries.

- When you share to ArcGIS Online, your data must be copied to ArcGIS Online.
- When you share to Portal for ArcGIS, you can choose Reference registered data or Copy all data.
  - To use the reference option, your data must be registered with the federated ArcGIS Server. Your data can be a file geodatabase, an enterprise database, or a folder containing your mosaic dataset. Data registration provides ArcGIS Server a list of locations with verified access.
  - You must choose the reference option if you want the edits to your source data to be dynamically displayed in the web layer, or if you want the edits to your web layer to be dynamically updated in your source data.

This tutorial

In this tutorial, you will share a raster tile layer, a vector tile layer, and optionally a map image layer to your ArcGIS Enterprise or ArcGIS Online, and then create a web app to compare the performances of these layers.

Data: A file geodatabase containing major earthquakes and hurricanes covering the US region between 2000 and 2008.

Requirements: Share the data as a raster tile layer, a vector tile layer, and a map image layer (optional), and display them side by side to compare their performances.

System requirements:
- ArcGIS Pro.
- ArcGIS Enterprise or ArcGIS Online (if you use ArcGIS Online, skip section 5.5, which is optional) and an account at the publisher or administrator level.

5.1 Add a portal connection in ArcGIS Pro

ArcGIS Pro needs a portal connection to either ArcGIS Online or Portal for ArcGIS. You can have connections to multiple portals, only one of which can be the active portal. Your active portal connection is your online workspace and the target to which you will share web layers.

If you have already added the connection to your portal and have set the connection as the active portal connection, skip this section.

1. Start ArcGIS Pro.

2. Click ⓘ About ArcGIS Pro in the lower-left corner. If you have opened a project, click the Project tab on the ribbon.

3. Click the Portals link.

4. Click Add Portal.

5. In the Add Portal dialog box, enter the URL to the portal you will use for this tutorial.

   • If you will use ArcGIS Online, specify **https://www.arcgis.com** or the URL to your ArcGIS Online for Organizations (the pattern is https://yourorg .maps.arcgis.com)
   • If you will use Portal for ArcGIS, the URL pattern is typically https:// portalhost.domain.com/portal. If you are not sure, ask your GIS administrator or instructor.

6. Click OK.

The connection is added to the portals list.

7. In the list of portals, right-click the connection. If you are not signed in, click Sign in and sign in with your account.

8. If you see a security alert that says, "Certificate Issuer for this site is untrusted or unknown," click Yes to proceed.

The alert message is typically because your Portal for ArcGIS uses a self-signed or domain-signed SSL/TSL certificate.

9. In the list of portals, right-click the connection you will use for the rest of the tutorial, and choose Set As Active Portal.

5.2 Author your map document

In this section, you will create a map document for use in sections 5.3, 5.4, and 5.5.

   Notes:
   - If you have not downloaded the sample data, navigate to **esri.com/gtkwebgis3**, and extract the files to C:\EsriPress.
   - This book does not address map design or cartography, so you will quickly design a simple map that will work for the tutorial.

1. Start ArcGIS Pro, and click Blank to create a new project.

2. For project name, specify **Natural_Disasters**, and click OK.

3. Click the Insert tab, and click the New Map button.

Project	Map	Insert	Analysis	View	Edit	Imagery	Share

New Map ▾ | 🔲 New Layout ▾   Import Map   Connections ▾ | 🔲 Toolbox ▾   Add Folder   Task ▾ | Bright Map Notes | Dark Map Notes | Light Map Notes | Paired Map Notes

Project | | | Layer Templates

4.   Click the Map tab, click the Basemap button, and select Light Gray Canvas.

5.   In the Catalog window, click the Project tab, right-click Folders, and choose Add Folder Connection.

Catalog   ▾ ⋈ ✕

Project | Portal | Favorites | History   ☰

⊕ 🏠 | *Search*   🔍 ▾

▷ 🔲 Maps

▷ 🔲 Toolboxes

▷ 🔲 Databases

▷ 🔲 Layouts

▷ 🔲 Styles

▷ 🔲 Folders   🔲 Add Folder Connection

▷ 🔲 Locators

6.   Click Computer > OSDisk (C:) > EsriPress, click the GTKWebGIS folder, and click OK.

A folder connection to this book's sample data is created, which allows you quick access.

7.   In the Catalog window, under the Project tab, look for Folders. Click the arrow next to the folder GTKWebGIS to expand the folder, expand the folder Chapter5, and expand Data.gdb. Notice there are feature classes.

8.   Drag the two layers (Earthquakes and Hurricanes) one by one to your map. They display with the default symbols.

9. In the upper-left quick access bar, click Save to save your project.

Next, you will style the Earthquakes layer based on magnitudes.

10. In the Contents pane, right-click the Earthquakes layer and click Symbology to open the Symbology pane.

11.   In the Symbology pane, select the following settings:

- For Symbology, choose Graduated Colors.
- For Field, choose Magnitude.
- For Color scheme, choose Yellow to Red.

Symbology   ▾ ⏸ ✕

Earthquakes   ☰

Symbology

Graduated Colors   ▾

Field   MAGNITUDE   ▾   ▦

Normalization   <None>   ▾

Method   Natural Breaks (Jenks)   ▾

Classes   5   ▾

Color scheme   [   ]   ▾

The new symbology appears on the map. Next, you will style the Hurricanes layer based on wind speeds.

12.   In the Contents pane, right-click the Hurricanes layer, and click Symbology.

13.   In the Symbology pane, select the following settings:

- For Symbology, choose Graduated colors.
- For Field, choose wmo_wind, the hurricane wind speed field.
- For Color scheme, choose Yellow to Red.

The Hurricanes layer symbols are updated on the map. Next, you will label the Earthquakes layer with magnitudes.

14.   In the Contents pane, right-click the Earthquakes layer, and click Labeling Properties.

15. In the Label Class pane under the Class tab, clear the Expression box. Double-click the Magnitude field.

16. Click the Position tab. If a Save Expression warning message occurs asking if you want to apply the expression, click OK. For Placement, select Top of point.

17. Click the Class tab and click Apply.

18. In the Contents pane, right-click the Earthquakes layer, and click Label.

The earthquakes are now labeled with magnitudes. Before you publish web layers, you typically must make sure your map is in Web Mercator, unless you have a good reason to use a different coordinate system.

19. In the Contents pane, double-click Map.

The Map Properties dialog box appears.

20. In the Map Properties dialog box, click the Coordinate Systems link on the left, and make sure the Current XY is WGS 1984 Web Mercator Auxiliary Sphere.

If the coordinate system is not Web Mercator, you can type in Web Mercator in the Search box to find and select the correct coordinate system.

21. Click OK to close the Map Properties dialog box.

22. On the upper-left quick access bar, click Save to save your project.

Now you have created a simple map document. Keep your map open if you are continuing with the next section.

5.3 Publish and use a vector tile layer

You can publish vector tile layers to either ArcGIS Online or Portal for ArcGIS. If you have Portal for ArcGIS, you should use it as the active portal for this whole tutorial, otherwise, you can publish to ArcGIS Online.

1. Continuing from the last section, click the Share tab, click the Web Layer list, and click Publish Web Layer.

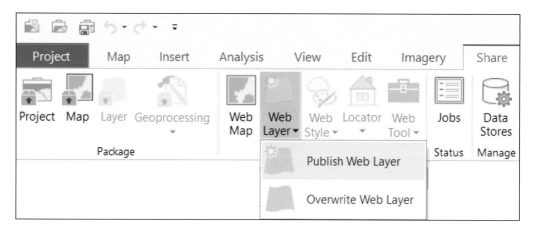

2. In the Share As Web Layer pane, under the General tab, select the following settings:

   - For Name, specify **NaturalDisasters_VectorTile**.
   - For Data, choose Copy all data (you won't see this option if your active portal is ArcGIS Online).
   - For Layer Type, click Vector Tile.
   - For Summary, specify **2000-2008 earthquakes and hurricanes**.
   - For Tags, specify **Earthquakes, Hurricanes**, and **GTKWebGIS**.
   - For Sharing Options, select Everyone.

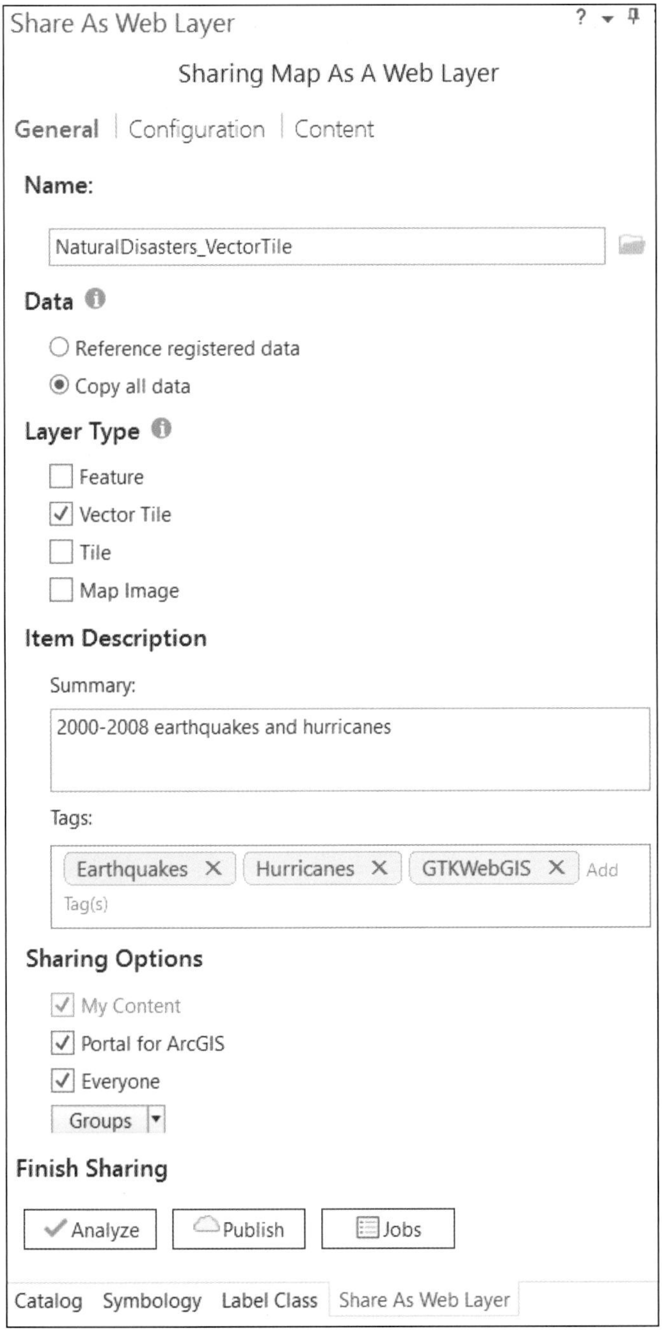

Share As Web Layer   ? ▾ ⚲

Sharing Map As A Web Layer

General | Configuration | Content

Name:

NaturalDisasters_VectorTile   🗄

Data ⓘ

○ Reference registered data
◉ Copy all data

Layer Type ⓘ

☐ Feature
☑ Vector Tile
☐ Tile
☐ Map Image

Item Description

Summary:

2000-2008 earthquakes and hurricanes

Tags:

| Earthquakes ✕ | Hurricanes ✕ | GTKWebGIS ✕ | Add
Tag(s)

Sharing Options

☑ My Content
☑ Portal for ArcGIS
☑ Everyone
Groups ▾

Finish Sharing

✔ Analyze   ☁ Publish   ▤ Jobs

Catalog Symbology Label Class Share As Web Layer

Next, you will define the tiling scheme.

3.   In the Share As Web Layer pane, click the Configuration tab.

4.   If your active portal is Portal for ArcGIS, click the Configure Web Layer Properties button ✎.

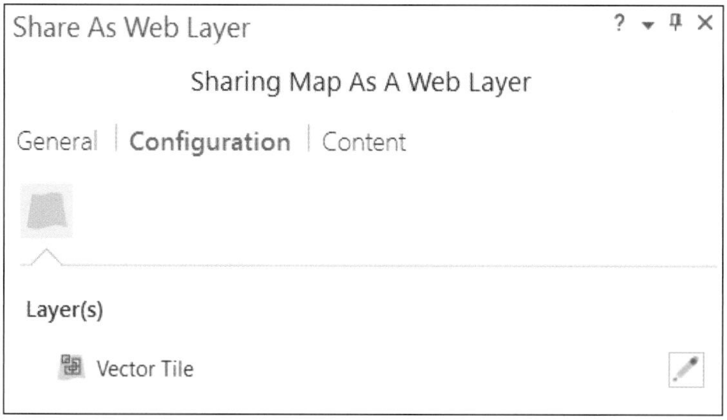

Share As Web Layer   ? ▾ ⊓ ✕

Sharing Map As A Web Layer

General | **Configuration** | Content

Layer(s)

   📇 Vector Tile   ✎

5.   Select the following settings:

- For Tiling Scheme, choose ArcGIS Online / Bing Maps / Google Maps
- For Levels of Detail, Use the slider and set it to 0 to 19.

The file size of vector tiles is typically much smaller than raster tiles. Also, because the tutorial data size is small, you can set the levels of detail to 0 to 19. If your dataset size is much larger, you may want to forego caching some of the extremely large or small scales in your tiling scheme.

- For Tiling Format, leave the option as Indexed.

The indexed option will have the tiles produced based on an index of feature density that optimizes the tile generation and file sizes. The Flat option will have the tiles generated for each level of detail without regard to feature density. This cache is larger than that produced with an indexed structure.

6. If your active portal is Portal for ArcGIS, in the upper-left corner of the Share As Web Layer pane, click the back button ⊖.

7. Click the Analyze button.

The results of the analysis appear on the Messages tab. You should see two warnings indicating that the basemap layers are not supported as vector tiles.

8. Click the 24078 warning message group to expand it.

9. Right-click the first warning message in the group, and choose Select Layer In The Contents Pane.

10. In the Contents pane, right-click the selected layer, and choose Remove.

11. Repeat steps 9 and 10 to remove the other basemap layer.

12. Click the Analyze button again. Notice there are no more errors and warnings.

13. Click the Save button to save your project.

Next, you will publish your vector tile layer.

14. Click the Publish button.

ArcGIS Pro will generate the vector tile package, upload the package to your active portal, and create a vector tile layer. The process may take some time if your data size is large and the network connection to your active portal is slow.

15. Once the web layer has been successfully published, click the Manage the web layer link.

Finish Sharing

| ✓ Analyze | ☁ Publish | ▤ Jobs |

✓ Successfully published web layer on 11/6/2017 12:07 AM
Manage the web layer   ✕

Catalog   Symbology   Label Class   Share As Web Layer

You will be directed to the details page of your web layer in your active portal. Next, you will use the vector tile layer as a basemap layer in a web map.

16. Click Open in Map Viewer.

The vector layer displays in the map viewer.

17. If you are not signed in, click Sign In and sign in with your portal account.

18. In the Contents pane, point to the vector tile layer, click the More Options button (note there are no options to enable pop-ups), and click Move to Basemap.

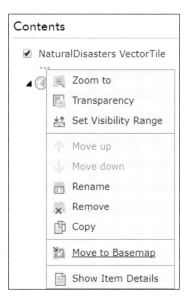

19. Zoom and pan the map to see the performance of the vector tile layer.

20. On the map viewer toolbar, click the Save button, select Save, set Title as **Natural disasters (vector tile)**, set the Tags as **Natural Disasters**, **GTKWebGIS**, and **Tutorial**, and click Save Map.

21. Click Share to share your web map with everyone.

In this section, you have published a vector tile layer and used it as a basemap in a web map. You will use this web map in section 5.6.

5.4 Publish and use a raster tile layer

The steps for this section are similar to the steps in the previous section. This section will be brief.

You can publish vector tile layers to either ArcGIS Online or Portal for ArcGIS. If you have Portal for ArcGIS, you should use it as the active portal for this whole tutorial, otherwise, you can publish to ArcGIS Online.

1. Continuing from the last section, in ArcGIS Pro, click the Share tab, click the Web Layer list, and click Publish Web Layer.

2. In the Share As Web Layer pane, click the General tab, and select the following settings:

    • For Name, specify **NaturalDisasters_RasterTile**.
    • For Data, choose Copy all data (you won't see this option if your active portal is ArcGIS Online).
    • For Layer Type, click Tile.
    • For Summary, leave it as **2000-2008 earthquakes and hurricanes**.
    • For Tags, leave them as **Earthquakes**, **Hurricanes**, and **GTKWebGIS**.
    • For Sharing Options, select Everyone.

Next, you will define the tiling scheme.

3. In the Share As Web Layer pane, click the Configure tab.

4. If your active portal is Portal for ArcGIS, click the Configuration tab and the Configure Web Layer Properties button ✎.

5. Make the following settings:

    • For Tiling Scheme, leave it as ArcGIS Online / Bing Maps / Google Maps.
    • Specify Levels of Details.
        ◦ Set it to 0 to 18, and notice the Estimated cache size is 41TB.
        ◦ Set it to 0 to 19, and notice the Estimated cache size is 164TB. Understand that the cache size grows four times for each additional level. Understand that the cache size is significantly larger than the vector tile layer you created in the last section.
        ◦ Set it to 0 to 7. Notice how the Estimated cache size is 10MB.

This cache size is a good size that can be published and cached quickly.

- Select Cache automatically on the server.

Share As Web Layer   ? ▾ ⊓

⊙   Tile Properties

❯ Caching

Tiling Scheme: ArcGIS Online / Bing Maps / Google Maps ▾ ⓘ

Levels of Detail

0   19

Minimum   Maximum
Level: 0   Level: 7
Scale: 1:591,657,528   Scale: 1:4,622,324

World   States / Provinces

Estimated cache size: 10MB

◉ Cache automatically on the server

Because the estimated cache size is small, it is easiest to build the entire cache automatically on the server. You have two other options:

- **Cache manually on the server:** If the cache is large, it is wise to build the cache manually in phases. You can manually build the cache from the ArcGIS Online website or Portal for ArcGIS website.
- **Cache locally:** Use this option if you want to minimize the number of credits consumed when you share to ArcGIS Online or if you want to reduce the number of resources used when you share to Portal for ArcGIS. When you use this option, the cache is temporarily saved as a tile package that is then uploaded and unpackaged automatically.

6. For you to finish this tutorial quickly, don't set the levels of detail beyond level 7.

Otherwise, you may have to wait for some time for the publishing and caching processes to complete.

7. If your active portal is Portal for ArcGIS, in the upper-left corner of the Share As Web Layer pane, click the back button ⊜.

8. Click Analyze. Fix any errors.

9. Click Publish.

The publishing process may take a couple of minutes depending on the network connection to your active portal.

10. Once the web layer has been successfully published, click the Jobs button to see the progress of the cache generation.

Job Status   ? ▾ ⊥ ✕

Job Queue   ☰

🔳 Publish NaturalDisasters_RasterTile
    Cache generation is in progress (50%)

11. Once the cache generation is completed, click the Share As Web Layer tab at the bottom, and click Manage the web layer.

Finish Sharing

[✓ Analyze]   [☁ Publish]   [▤ Jobs]

✓   Successfully published web layer on 11/6/2017 12:40
     AM. View cache status in the Job Status pane.   ✕
     Manage the web layer

Catalog   Symbology   Label Class   Share As Web Layer   Job Status

You are directed to the details page of your web layer. Next, you will use the layer as a basemap layer in a web map.

12. Click Open in Map Viewer.

13. If you are not signed in, click Sign In, and sign in with your account.

14. In the Contents pane, point to the raster tile layer, click the More Options button, note there are no options to enable pop-ups, and click Move to Basemap.

15. Zoom in and notice the tile layer disappears once the zoom level goes beyond 7.

16. Zoom and pan the map. Notice the performance of the raster tile layer.

17. On the map viewer toolbar, click the Save button, select Save, set Title as **Natural disasters (raster tile)**, set the Tags as **Natural Disasters**, **GTKWebGIS**, and **Tutorial**, and click Save Map.

18. Click Share to share your web map with everyone.

In this section, you have published a raster tile layer and used it as a basemap in a web map. You will use this web map in section 5.6.

5.5 Publish and use a map image layer (optional)

Note: You can only publish map image layers to ArcGIS Enterprise. If you don't have ArcGIS Enterprise, skip this section.

The steps for this section are very similar to the steps in the previous section. This section will be brief.

1. Continuing from the last section, in ArcGIS Pro, click the Share tab, click the Web Layer list, and click Publish Web Layer.

2. In the Share As Web Layer pane, under the General tab, select the following settings:

   - For Name, specify **NaturalDisasters_MapImage**.
   - For Data, select Copy all data.
   - For Layer Type, click Map Image.
   - For Summary, leave it as 2000-2008 earthquakes and hurricanes.
   - For Tags, leave them as Earthquakes, Hurricanes, and GTKWebGIS.
   - For Sharing Options, select Everyone.

3. In the Share As Web Layer pane, click the Configuration tab.

4. Click the Configure Web Layer Properties button ✎ for Map Image. Notice that map image layers support Map, Data, and Query operations. Also notice the options to draw this map image layer dynamically from data or using tiles from a cache.

5. In the Share As Web Layer pane, click the back button in the upper-left corner.

6. Click Analyze. Fix any errors at least, if not the warnings.

7. Click Publish.

Your layer should be published faster than the tile layers. No tile generation is needed.

8. Once the web layer has been successfully published, click Manage the web layer.

You will be directed to the details page of your web layer. Next, you will use the layer in a web map.

9. Click Open in Map Viewer.

10. If you are not signed in, click Sign In, and sign in with your portal account.

11. Zoom and pan the map. Understand that dynamic map image layers display slower than tile layers.

12. On the map, click Earthquakes and Hurricanes, and notice that pop-ups appear.

13. On the map viewer toolbar, click the Save button, select Save, set Title as **Natural disasters (map image)**, set the Tags as **Natural Disasters, GTKWebGIS**, and **Tutorial**, and click Save Map.

14. Click Share to share your web map with everyone.

In this section, you have published a dynamic map image layer and used it in a web map. You will use this web map in section 5.6.

5.6 Compare map layers side by side

ArcGIS Compare Analysis is a configurable app template with the ability to display and compare up to four web maps at a time. The first web map chosen in the app controls the extent of the succeeding web maps. In this section, you will use this template to create a side-by-side comparison web app to display the web layers you published in the previous sections. If you don't have ArcGIS Enterprise and skipped section 5.5, you can still work on this section.

1. In a web browser, go to the active portal you used in previous chapters, and sign in.

2. Click the Content tab. Under My Content, click Create and choose App > Using a Template.

3. In the Create a New Web App window, click Compare Maps/Layers on the left.

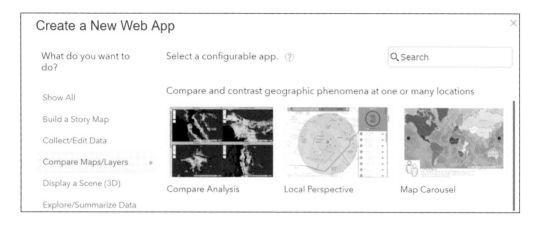

4. Click the Compare Analysis template, and click Create Web App.

5. In the Create a New Web App window, set the Title as **Compare map layers**, set Tags as **GTKWebGIS** and **Tutorial**, click Done.

You will be directed to the configuration page of the app.

6. In the Select Web Map(s) window, click My Content, select the web maps you created in the previous sections.

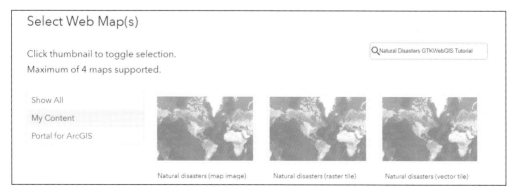

If you have many web maps, you can search for the Natural Disasters GTKWebGIS Tutorial to narrow down the web maps.

If you don't have ArcGIS Enterprise and skipped the last section, you won't see the map image web map. It's okay to skip this web map.

7. In the results, select the web maps you created in the previous sections, and click OK.

8. For side panels, unselect Include side panel and unselect Show side panel.

9. Click Save and click Close.

You will be directed to the app's details page.

10. Click Share, share the app with everyone or certain groups, and click OK.

11. Click View Application.

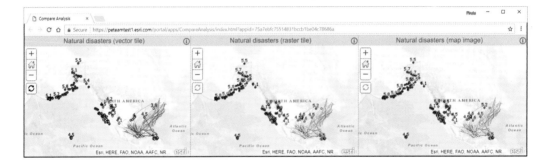

You should see the web maps display side by side. If you pan and zoom the first map, you'll notice the rest of the maps will follow. The tutorial data size is small. The performance differences of the web maps may not be dramatic, but you should notice that the vector tile layer displays faster than the raster tile layer, which displays faster than the map image layer. This observation is generally true, though you may occasionally see different results.

QUESTIONS AND ANSWERS

1. Why should I project the data that I want to share as web layers into Web Mercator?

   Answer: Unless your study area lies within the northern or southern polar regions or faces some specific requirements, your map service will most likely be displayed on top of other basemaps, such as the basemaps in ArcGIS Online. For that reason, you should project your data to Web Mercator first; projecting the data first prevents ArcGIS Enterprise and ArcGIS Online from having to project your data dynamically (on the fly). Projecting your data to Web Mercator, which you can do in ArcGIS Pro, can improve the performance of both the map service and the web app.

2. After publishing my web layers from ArcGIS Pro, I updated my GIS data. Will my data updates appear in my web layers automatically?

   Answer: It depends. Web layers published to ArcGIS Online won't pick up your local data updates. Feature layers and dynamic map image layers published to ArcGIS Enterprise using referencing registered data will reflect your date updates.
   You can register your data folders and databases with ArcGIS Server using ArcGIS Server Manager or using ArcGIS Pro (in the Share As Web Layer pane, select Reference registered data, click Analyze, and in the results, double-click the message 00231: Layer's data source must be registered with the server).

3. I created a web app using the web layers that I published to my university's ArcGIS Enterprise. However, I can view this web app only when I am using the computers at my university campus. Why?

   Answer: The ArcGIS Enterprise is installed on a private server at your university and is accessible only through the internal network. For some situations, this limitation means enhanced security. For other situations, this limitation can be inconvenient.
   If you want to make your web layers, maps, and apps accessible off campus, try one of the following solutions:
   - Use Virtual Private Network (VPN): If your university allows VPN access, you can use VPN to get into your university network, even when you are off campus.
   - Use a reverse proxy: A reverse proxy is a type of proxy server that retrieves resources on behalf of a client. Your university system administrator can set this proxy up.

- Have your university's ArcGIS Enterprise installed on a public server that is accessible to users via HTTP or HTTPS from the external network.

4. This chapter mentioned that the labels of vector tile layers can remain upward as the map rotates. Is there a live demo where I can test this feature?

    Answer: Yes. In a Chrome web browser, open C:\EsriPress\GTKWebGIS\ Chapter5\displayWebMap.html, right-click the map to rotate, and notice the labels of the earthquakes remain upward.

ASSIGNMENT

Assignment 5: Publish a raster tile layer and a vector tile layer, and compare the layers side by side in a web app.

Data: The C:\EsriPress\GTKWebGIS\Chapter5\Assignment.gdb has the following two layers:

- Wildfires in 2015 and 2016 with burned areas greater than or equal to 10 acres. Refer to firehistory.xml for the metadata
- Tornado paths in 2015 and 2016. Refer to SPC_severe_database_description .pdf for meanings of the attribute field names

System requirements: You can publish to either ArcGIS Online or ArcGIS Enterprise.

Tips:

- You may style the wildfires on the TotalAcres field.
- You may style the Tornadoes layer on the Magnitude field.
- Be careful with the size of the raster tiles that will be generated. Keep the size small (smaller than 40MB for instance) so that you can complete your homework quickly and the tiles won't take too much space in your portal.

What to submit:

- URLs to the item details pages of your web layers
- URL to your web app
- A paragraph summarizing the differences between raster tile layers and vector tile layers

Resources

"About using your portal with ArcGIS Server," http://server.arcgis.com/en/portal/latest/administer/windows/about-using-your-server-with-portal-for-arcgis.htm (or http://arcg.is/2zny7LL).

"ArcGIS Enterprise: An Introduction," Shannon Kalisky and Thomas Edghill, https://www.youtube.com/watch?v=ZcBh3M-uQNw (or http://bit.ly/2yDa7VK).

"ArcGIS Enterprise: Configuring a Base Deployment," https://www.esri.com/training/catalog/59c40dcade53ed5705e39a67/arcgis-enterprise:-configuring-a-base-deployment (or http://arcg.is/2zoqJ0u).

"ArcGIS Enterprise—Improving Installation and Configuration," Julia Guard, https://www.youtube.com/watch?v=_Cp4SD7_2wg (or http://bit.ly/2Ap9Se9).

"ArcGIS Enterprise Portal Collaboration," Kevin Kelly and Betsy Leis, https://www.youtube.com/watch?v=bOzO0MCaBLU (or http://bit.ly/2hgDLbN).

"Creating Vector Tiles in ArcGIS Pro," https://www.esri.com/training/catalog/5851736fd33f8b0b47b78e26/creating-vector-tiles-in-arcgis-pro (or http://arcg.is/2AdYIIj).

"Explore the World of ArcGIS Enterprise," https://www.esri.com/training/catalog/599c786d8907337d57562b13/explore-the-world-of-arcgis-enterprise (or http://arcg.is/2vgCPpM).

"Introduction to sharing web layers," http://pro.arcgis.com/en/pro-app/help/sharing/overview/introduction-to-sharing-web-layers.htm (or http://arcg.is/2jamH88).

"Map image layer," http://pro.arcgis.com/en/pro-app/help/sharing/overview/map-image-layer.htm (or http://arcg.is/2zidQYl).

"Migrating from ArcMap to ArcGIS Pro," https://www.esri.com/training/catalog/59161f89dfcdee7bac39d5b3/migrating-from-arcmap-to-arcgis-pro (or http://arcg.is/2zk65kK).

"Sharing Maps and Layers with ArcGIS Pro," https://www.esri.com/training/catalog/57630436851d31e02a43f0eb/sharing-maps-and-layers-with-arcgis-pro (or http://arcg.is/2xNWbqe).

"Sharing Web Layers and Services in the ArcGIS Platform," Melanie Summers and Derek Law, https://www.youtube.com/watch?v=tWPSzvXbsfc (or http://bit.ly/2znkSus).

"Tile layers," http://doc.arcgis.com/en/arcgis-online/reference/tile-layers.htm (or http://arcg.is/2zCDA2i).

"Vector tile layer," http://pro.arcgis.com/en/pro-app/help/sharing/overview/vector-tile-layer.htm (or http://arcg.is/2hc6NWB).

"Web tile layer," http://pro.arcgis.com/en/pro-app/help/sharing/overview/web-tile-layer.htm (or http://arcg.is/2hNTs7c).

"What is ArcGIS Enterprise?" http://server.arcgis.com/en/server/latest/get-started/windows/what-is-arcgis-enterprise-.htm (or http://arcg.is/2hbrUbi).

1
2
3
4
5
6
7
8
9
10

CHAPTER 6

Spatial temporal data and real-time GIS

Everything that happens, happens somewhere and sometime. Real-time GIS has been used to handle objects and events that move, appear, and change through time. Today, mobile phones, sensor networks, smart cities, and the IoT generate large volumes of real-time data at high velocities. Such real-time data demands real-time GIS to locate and track targets, as well as store, manage, search, display, and analyze the data. The overview section of this chapter introduces the basic concepts of spatial temporal data, the values and challenges of IoT and smart cities, and the ArcGIS products that can meet these challenges. The ArcGIS products introduced include ArcGIS® GeoEvent™ Server and Operations Dashboard for ArcGIS®. The tutorial sections in this chapter teach how to use real-time layers in web maps, create dashboard apps, create time-enabled layers, and animate time-series data.

Learning objectives

- *Understand spatial temporal data concepts and terminology.*
- *Learn about IoT, sensor networks, smart cities, and other related frontiers.*
- *Understand the advantages of stream services.*
- *Learn about ArcGIS GeoEvent Server capabilities.*
- *Use Operations Dashboard for ArcGIS to monitor real-time data.*
- *Create time-enabled web layers and apps to animate time-series data.*

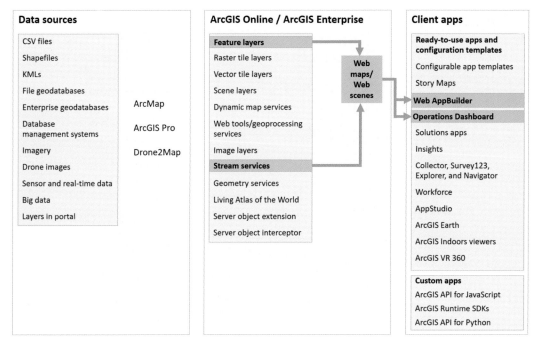

The ArcGIS platform offers many Web GIS apps and many ways to build apps. The green lines in the figure highlight the technology presented in this chapter.

Spatial temporal data and real-time GIS basics

Time is an important dimension of GIS data. Temporal data include observations of objects and events that move or change through time, such as when or where an observation took place and what activity was observed. Spatial temporal data can be categorized into the following groups:

- Moving (for example, live feeds of airplanes, buses, cars, ambulances, fire engines, and trains)
- Discrete (for example, criminal incidents, earthquakes, and geotagged Twitter and Instagram feeds)
- Stationary (for example, wind speed and direction measurements at weather stations, water levels at stream gauges, highway and street traffic speed, and live pictures and videos from surveillance cameras)
- Change (for example, perimeters of wildfires, flooded areas, urban sprawl, and land use and land cover changes)

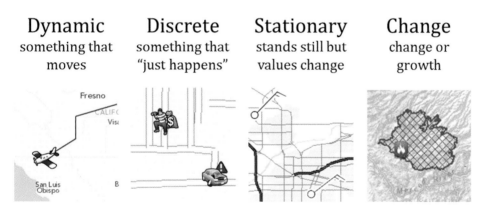

Dynamic	Discrete	Stationary	Change
something that moves	something that "just happens"	stands still but values change	change or growth

Types of spatial temporal data.

Real-time GIS refers to GIS that handles current and continuous data, which can be the latest position, altitude, speed, direction, temperature, pressure, concentration, or water level of various sensors and other objects. Real-time GIS can provide better situational awareness, enhance emergency response, and support better decision making by delivering data and performing analysis when events happen.

In spatial temporal GIS data, the time value of an event can be a point in time or a duration of time.

- **Point in time:** One example is the time a lightning strike happened. Such time values are typically stored in a single attribute field.
- **Duration of time:** An example is the time a wildfire started and ended. Such time values are typically stored in two fields, one for the start time, and the other for the end time.

You will encounter some key terms when working with spatial temporal data, which follow.

- **Time measurement systems (in other words, units):** Time can be expressed in many units such as in years, months, days, hours, minutes, and seconds.
- **Time reference systems (in other words, time zones and rules for daylight savings time):** The most often used time zones are Greenwich Mean Time (GMT) and Coordinated Universal Time (UTC). Both reference the time at the Prime Meridian (0 longitude), but GMT accounts for daylight savings time while UTC doesn't. To avoid confusion about time zones and daylight-savings changes, most information systems, including ArcGIS, store dates as the seconds or milliseconds from January 1, 1970, in UTC.
- **Time representations:** Time can be represented in many different formats and languages (for example, 12/18/2018, December 18, 2018, 18/12/2018, 18 December 2018). ArcGIS and most information systems support flexible date and time representations based on the specified formats, browser locales, and time zones.
- **Temporal resolution:** Temporal resolution typically refers to the time interval at which events are sampled. For instance, an Automatic Vehicle Location (AVL) system checks the locations of ground vehicles every 15 seconds, and a weather station reports temperatures every 15 minutes. Smaller temporal intervals will result in larger data sizes being transmitted, ingested, and stored, while larger temporal intervals will likely result in lower temporal resolution.

IoT

Spatial temporal data come from many sources, ranging from manual data entry to data collected using observational sensors or generated from simulation models. Today, the advent, maturity, and affordability of various sensor technologies have brought about and amplified spatial temporal data. The proliferation of smartphones and tablets have accelerated the explosion of VGI. Improved accuracy of GPS, high-speed wireless data communication, and affordable sensors have propelled the IoT.

IoT is the network of physical objects or "things" embedded with sensors and network connectivity, which enable these objects or "things" to collect and exchange data. "Things." in the IoT sense, refer to a variety of devices such as airplanes, taxis, bicycles, lights, refrigerators, sprinklers, heart monitoring implants, biochips, security cameras, and unmanned automobiles. The notion of the IoT is being adopted rapidly across the globe. Experts estimate that the IoT will consist of about 30 billion objects by 2020, and the global market value of the IoT will reach $7.1 trillion by 2020.

The IoT is a giant network of connected "things" and people.

In recent years, the science community as well as the government and private sector communities have embraced the concepts of the IoT to support the creation of systems and products ranging from enterprise applications to consumer applications.

- Enterprise IoT applications include smart cities, infrastructure management, environment quality monitoring, smart retail-inventory management, and precise agriculture.
- Consumer IoT applications include connected cars, connected health, and smart homes.

Sensor data collected in the IoT ecosystem require context to be understood and valuable. Geolocation provides that context, and GIS can transform the raw data into useful information that is, ultimately, actionable intelligence. For example, smart cars can share information with other vehicles to warn them about slippery road conditions. Warning other cars about the hazard requires knowing where the hazard is as well as the other vehicles' proximity to the hazard. Only by understanding the location of the slippery road segment can other vehicles determine the relevance of the warning to their planned routes. Location and geoanalytics transform this sensor data into actionable intelligence.

In the concept of the IoT, connected vehicles can exchange information about road conditions with adjacent vehicles directly or through the cloud.

Smart city

An important application area of the IoT is smart cities. A smart city uses IoT devices, such as connected sensors, to supply information that will assist the city in managing assets and resources efficiently. The smart city concept integrates IoT technology to make more efficient use of physical infrastructure, as well as engage effectively with citizens and decision makers so they can learn, adapt, innovate, and respond more promptly to changing circumstances.

Implementing smart cities is not just deploying sensors around cities and collecting the sensor data. Without the means to visualize and act on the data, the power of the data is muted. Location data plus time stamps give city administrators the capability to know where and when something is, or was, and can transform raw sensor data into actionable information. Implementing smart cities needs GIS to ingest the data, store the data, and perform spatial temporal analysis on the data. The analysis can detect patterns and trends; predict what might happen in the future; and make cities more sustainable, secure, and enjoyable.

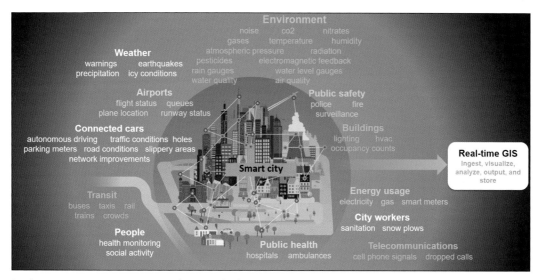

Implementing a smart city is more than just deploying sensors. Ingesting and processing the spatial temporal data is critical for the success of smart city initiatives.

Smart home

A smart home, also known as a smart house or home automation, is a consumer application of the IoT that automates various components of a home. Smart home implementations often involve Wi-Fi-connected smart sensors or controllers, such as smart thermostats, and an artificial intelligence controlling device, such as Amazon Echo. The immediate benefits of a smart home include giving users the ability to remotely manage the lighting, heating and air-conditioning, media, and security systems in their homes. Long-term benefits include creating a more environmentally friendly home through the automation of some of these functions.

The development of smart homes is still in its infancy, but GIS will revolutionize this development. For example, smart sprinkler systems can connect to location-based weather forecasts, understand the upcoming weather in your city, and automatically adjust the watering time based on the weather forecast. Another example is the smart thermostat in your home today, which allows you to set the home temperature based on the time of day and adjust it both remotely and manually. But let's say you decide to come home early one day, and you didn't adjust your home temperature manually. The thermostat will typically not have advanced warning of your early home arrival, so the temperature when you get home could be too cold or too hot. But now imagine your smart thermostat is connected to your smartphone, which is connected to a smart home GIS app. You can use the GIS app to erect a Geofence around your workplace, so if you leave work early, your phone will trigger the Geofence around your workplace and the app will alert your thermostat to set the temperature to your desired comfort level before you get home.

Web GIS technologies for real-time GIS

The fast development of the IoT demands real-time GIS to locate and track "things"; store, manage, search, display, and analyze the data; and make predictions. On the other hand, the ubiquitous IoT sensors challenge real-time GIS not just by the volume of data to ingest and process, but also by the velocity of the data to ingest and process.

ArcGIS GeoEvent Server

GeoEvent Server is the core of Esri's real-time GIS technology. The server can connect numerous types of streaming data, perform continuous data processing and analysis, and send updates and alerts when specified conditions occur, all in real time. GeoEvent Server is scalable and can meet the requirements to ingest, process, and store high-volume and high-velocity real-time data generated by the IoT.

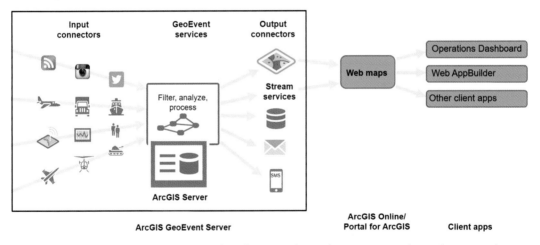

GeoEvent Server provides input connectors for taking in real-time data, processors for performing real-time filtering and analysis, and output connectors for generating results, which further support web maps, web apps, and big data analysis.

GeoEvent Server has the following four main categories of capabilities:
- **Ingest:** This component interfaces with various data sources. It provides ways to communicate directly with individual sensors or sensor networks, social network feeds, and other real-time data streams by various communication transport protocols. Examples of transport protocols include Transmission Control Protocol (TCP), User Datagram Protocol (UDP), HTTP, REST, File Transfer Protocol (FTP), file, email, Web Socket, and more. The ingestion component also provides the means to understand and translate different data formats (for example text CSV, JSON, XML, binary, and so on) into the Web GIS internal data structure.

- **Process:** This component processes the real-time data received and translated by the ingestion subsystem. This ingestion subsystem provides filters and processors.
  - Filters remove events that do not satisfy specified criteria. Filters are generally attribute filters, spatial filters, or a combination of both. An attribute filter can be configured with one or more attribute expressions (for example, the altitude of an airplane is less than 50 meters and the speed is greater than 100 miles per hour). Spatial filters filter events based on their spatial relationships with a Geofence (for example, when an airplane is entering and exiting a storm; when a delivery truck is approaching a customer's home).
  - Processors perform specific actions, such as creating a buffer around an earthquake, geoenriching the buffer to find out the affected demographics and calculating the estimated property damages.

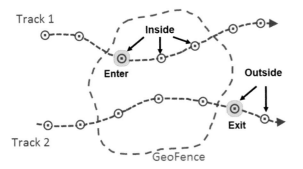

GeoEvent spatial filters can detect when an object enters or exits a Geofence and when an object is inside or outside the Geofence. These events can be configured to trigger certain alerts or actions.

- **Output:** The output component has similar characteristics to the ingest component but in the opposite direction. This component manages the connections with web clients, distributes data via various transport protocols, and sends various alerts or updates. The output subsystem can send data to a data store; send data to web clients; send alerts via emails, SMS, or as tweets; or send specific commands to activate external systems to perform certain tasks.
- **Archive:** ArcGIS supports various conventional and newer database technologies. The conventional relational database technology has been designed for data that don't change at very high rates or volumes. Experience has shown that a single instance of a relational database has a typical limit of 300 records per second on writing and updating. Therefore, newer technologies are needed to meet the challenge of the IoT. The ArcGIS Data Store, a component of ArcGIS Enterprise, provides the spatiotemporal configuration type that is designed specifically to handle spatial temporal data at high speeds and high volumes. The ArcGIS Data Store provides increased scalability by supporting node and cluster-type deployment patterns. ArcGIS can also work with other newer database technology, such as NoSQL, MongoDB, and Hadoop, to handle the high-frequency writing and retrieving of big data.

Steps to create GeoEvent services

A GeoEvent service configures the flow of events, what input connectors to accept and transform the data, what filtering and processing steps to perform, and what output connectors to send the results. You can use ArcGIS GeoEvent Manager to create and manage GeoEvent services.

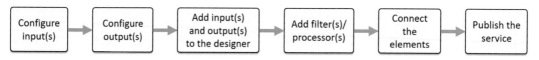

Steps to create GeoEvent services.

Here are the general steps to create a GeoEvent service:

- Configure the inputs and outputs using the existing connectors or custom connectors that you can download or develop. If you archive the real-time data using ArcGIS Data Store, MongoDB, or Hadoop, you must register the data stores in the ArcGIS Server Manager.
- Add the inputs and outputs to the GeoEvent service designer.
- Add filters and processors to the designer, and configure them as needed.
- Connect the inputs, filters, processors, and output elements.
- Publish the service.

GeoEvent Manager provides a simple designer interface similar to ArcGIS ModelBuilder. The designer allows you to add inputs and outputs, connect them with processors and/or filters, and publish your GeoEvent service.

Deliver real-time data from servers to clients

ArcGIS supports both the poll and push ways to deliver real-time data.

- **Poll** is the traditional approach in which a client periodically (for example, every 30 seconds) polls the server to retrieve the latest data. For example, in ArcGIS map viewer, you can configure the layers in your web map to refresh at an interval between 6 seconds and 1 day.
- **Push** is a new way to serve data in basically real time by using the HTML5 WebSocket protocol. For instance, GeoEvent stream services can push data to the client side. Stream services are especially useful for visualizing real-time data feeds that have high data volumes or data that change at unknown intervals. In ArcGIS map viewer, you can add a stream service to your web maps and have the controls available to you to start and stop the data stream.

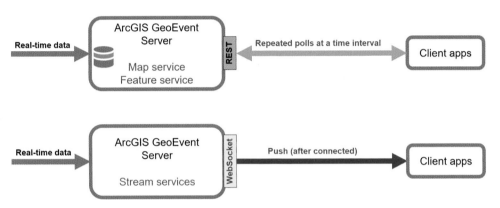

Using the poll technology (top), real-time GIS data is first saved in a feature class and then exposed as a feature or map service for the client to poll periodically. Using push technology (below), the real-time GIS data is pushed out to the client immediately as stream services via a WebSocket.

ArcGIS map viewer can take real-time data layers via poll and push. Web maps with real-time layers can be transformed into web apps using configurable apps, Web AppBuilder for ArcGIS, Operations Dashboard for ArcGIS, or custom JavaScript code.

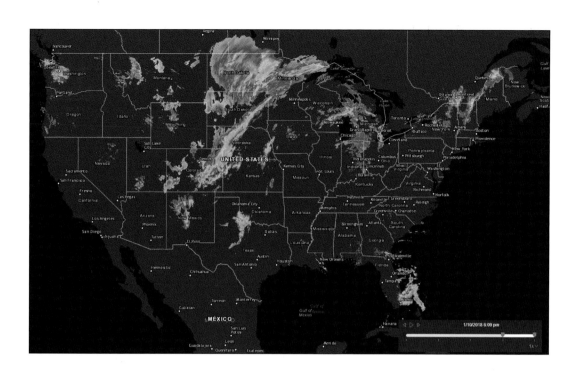

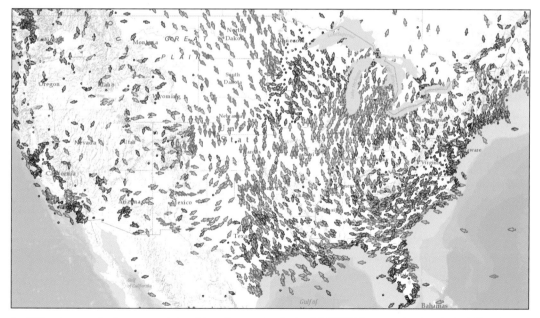

ArcGIS web apps display near real-time Next-Generation Radar (NEXRAD) imagery (top) and wind directions (bottom).

Operations Dashboard for ArcGIS

When you drive a car, you rely on the car's dashboard to monitor conditions such as the current speed, fuel level, gearshift position, and engine status. Similarly, you need a dashboard to gain situational awareness when you are responding to emergencies, managing workforces, and performing other operations. Esri provides Operations Dashboard for ArcGIS for such needs. Operations Dashboard provides a common view of the systems and resources you manage. The dashboard integrates web maps and a variety of data sources to create comprehensive operational views. The views include charts, lists, gauges, and indicators, which update automatically as the underlying data change. These live views allow you to monitor and track live events in real or near real time.

Dashboards with real-time data can be applied in many powerful ways:

- Law enforcement agencies monitor crimes as they are reported, as well as incoming 911 calls.
- Environmental protection agencies monitor air-quality and water-quality sensor data.
- Companies use real-time social media feeds, such as Twitter, to gauge feedback and monitor social sentiment about particular issues.

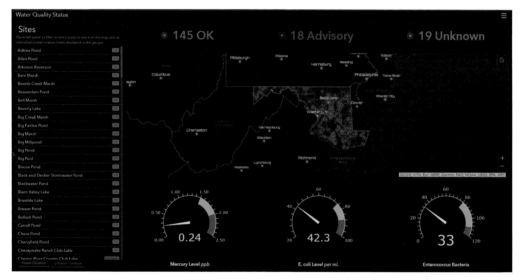

A water-quality dashboard provides the live view of various chemical sensor data in maps, lists, gauges, and indicators.

Operations Dashboard has a Windows edition and two web browser editions. The Windows edition can create operation views and display multiple views. An older browser edition is for displaying the single-display operation views created by the Windows edition. A new browser edition can create and display dashboards. The tutorial section of this chapter will introduce the new browser edition.

Animate time-series data

You can play back time-series data using maps and charts, even with animation effects. By animating time-series data, you can visualize the data at each step and see patterns and trends emerging over time. With animation, you can provide the following capabilities:

- Display discrete events, such as crimes, accidents, and diseases over time.
- Visualize the value change at stationary objects, such as air-quality sensors and weather stations.
- Map the progression of a wildfire, flood, or land use and environmental change over time.
- Replay the events of an emergency, review when and how different departments responded, and learn from the past to improve emergency management and responses.

To animate time-series data using ArcGIS, you create a time-enabled layer using either ArcGIS Desktop or ArcGIS Online, add the layer to a web map, and create a web app using an ArcGIS ready-to-use web client.

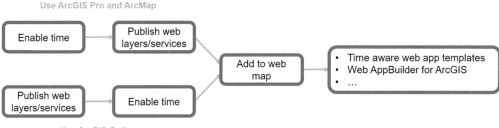

Typical workflow to create web apps that can animate time-series data.

- Publish time-enabled web layers: You can publish time-series data to ArcGIS Online or ArcGIS Enterprise as time-enabled web layers. To meet the data requirement for time-enabled layers, each feature must have a single date attribute field or two date fields (in other words, the beginning and ending date fields). If you use ArcGIS Online or Portal for ArcGIS to publish hosted layers, the date fields must be date type. If you use ArcGIS Desktop to publish your layer, the date fields can be date type or other data types, as long as you can express their values in "sortable" formats, such as YYYY, YYYYMMDD, string, or number format.

Layer Properties: US_Cities   ✕

General	Layer Time	Each feature has a single time field ▾
Metadata		
Source	Time Field	Year ▾
Elevation		
Selection	Time Format	YYYY ▾
Display	Time Extent	1/1/1790 ▦ - 1/1/2000 ▦
Cache		
Definition Query		Calculate
Time		☐ Data is a live feed
Range		Rate [1] Seconds ▾
Indexes	Time Zone	<None> ▾
Joins		
Relates		☐ Adjust For Daylight Saving
Page Query		Learn more about time properties

OK   Cancel

In ArcGIS Pro and ArcMap, you can open the properties window of a layer, and specify the time field(s). This action will enable time on this layer. If you share this layer to ArcGIS Online or ArcGIS Enterprise, you will create a time-enabled web layer.

- Add the time-enabled layer to a web map: Optionally, you can configure the time settings such as play speed, start time, end time, and time step. You can also specify the time interval in which the data is shown on the web map as time passes (for example, the earthquakes in each month, or progressively display all the data; the earthquakes from the starting time to the current time interval).
- Create a web app from the web map: You can create a web app by using an ArcGIS web app template that is time aware, or by using Web AppBuilder for ArcGIS. These apps provide a time slider or time widget for your users to animate the data.

This tutorial

This tutorial includes two case studies.

Case Study 1 (sections 6.1–6.3): Create a dashboard web app for the City of Redlands to coordinate response efforts for the city's medical, crime, and fire departments.

Functional requirements: The dashboard web app must have the following capabilities:

- Monitor the near-real-time locations of police, fire, and medical vehicles.
- Monitor the near-real-time locations and details of the incidents reported from 911 calls.
- List the most recent incident details.
- Display the counts, categories, and statuses of incidents in easy-to-understand charts.

Data: Provided as web layers.

System requirements:

- An ArcGIS Online administrator- or publisher-level account.
- Chrome web browser for inspecting the poll and push traffic.

Case Study 2 (sections 6.4–6.5): Create a web app to illustrate the spatial and temporal patterns of population change in major US cities over the last two centuries.

Functional requirements: Your web app must illustrate the spatial and temporal patterns of population change with animations.

Data: Provided as a CSV file.

System requirements:

- An ArcGIS Online or ArcGIS Enterprise administrator- or publisher-level account.

6.1 Create a web map with real-time layers

1. Open the Chrome web browser, go to ArcGIS Online (**http://www.arcgis.com**) and sign in.

2. Search for Redlands Dashboard Map owner:GTKWebGIS, and unselect the Only search in your organization option.

3. Click the thumbnail of the found web map to open it in the map viewer.

The web map has the incidents, police, fire, ambulances, facilities, and traffic service layers. These layers are configured to refresh regularly to display the latest data.

4. In the Contents pane, point to the Police layer, click the More Options button, select Refresh Interval, and notice this layer refreshes every 0.1 minutes.

This layer is polled every 6 seconds, and police cars may move on the map if the new locations are changed.

5. Repeat the previous step to review the refresh interval of all the other layers.

The Emergency Facilities layer and the CalTrans Cameras Redlands layer have no refresh intervals because these layers don't change. Other layers have intervals that range from 0.1 minute to 5 minutes.

Next, you will add two additional temporal layers.

6. On the map viewer tool bar, click the Add button and choose Search for Layers.

7. In the Search for Layers pane, search for "dashboard sample data owner:GTKWebGIS" in ArcGIS Online. Clear the Within map area box.

There are two layers in the search results. CalTrans_Cameras_Redlands is a simplified layer showing the live cameras by the California Department of Transportation (CalTrans). Helicopters is a stream service showing the real-time locations of the city's police helicopters. The helicopter data is simulated.

8. In the list of search results, click Add next to each of the layers to add the layers to the map.

9. Click Done Adding Layers.

10. On the map, click a few CalTrans cameras to view the new real-time highway photos in the pop-ups.

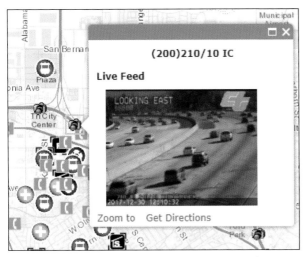

The photos in the pop-ups are captured every few minutes, as you can tell from the time stamps on the photos. Live videos of the cameras are available at CalTrans's websites.

Next, you will inspect the poll and the push ways to receive real-time or near-real-time data.

11. In the Contents pane, point to the Helicopters layer, click the More Options button, select Streaming controls, click Stop streaming, and click Start streaming.

Contents

☑ Helicopters

☑ CalTrans Camer   ▣ Transparency
                     ⚖ Set Visibility Range

                     🖳 Configure Pop-up

                     📵 Streaming controls   Stop streaming
                                                 Limit observations to current map area
                     📄 Show Item Details   Clear previous observations
                     💾 Save Layer

12. Press the F12 key on your keyboard to bring up the Developer tools, or click the Chrome options button, select More Tools > Developer tools.

13. In Developer Tools, click the Network tab, notice there are new query requests coming up.

These are the poll requests to refresh the incidents, police, fire, and ambulances layers every 6 seconds.

14. In the Filter input box, specify Stream, and click the found subscribe request.

15. Click the Frames tab, click a message, and review the actual data received.

The subscribe request initializes a WebSocket handshake with the server. After the handshake is successfully made, data streams come in as soon as the server has newer data.

16. Close the Developer Tools window.

Next, you will style the Helicopters layer.

17. In the Contents pane, point to the Helicopters layer, and click the Change Style button.

18. Click Symbols, click Use an Image, set the URL as **https://i.imgur.com/ LhzIJ82.gif**, click the plus (+) button, set Symbols Size as 24 px, and click OK.

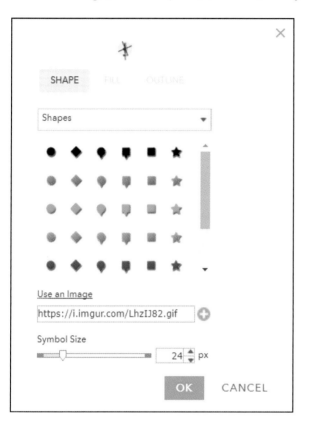

The symbol uses an animated GIF to represent flying helicopters. The symbol heads north by default. Next, you will rotate the symbol based on helicopter flight directions.

19. Click Rotate symbols, select the heading field, and leave Clockwise from 12 selected.

> ☑ Rotate symbols (degrees)
>
> Heading   ▼
>
> ◉ Clockwise from 12
> ○ Counterclockwise from 3

20. Click OK.

21. Notice a helicopter with rotating blades on the map.

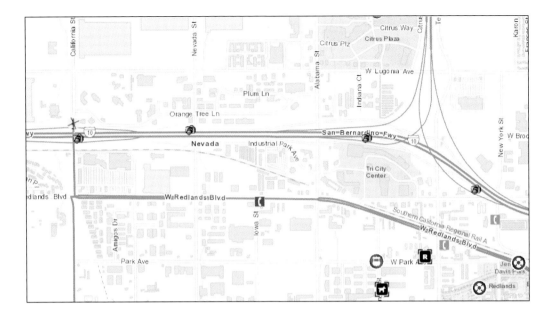

22. On the Map Viewer toolbar, click Save as to save your web map.

23. Click the Share button to share your web map with Everyone.

In this section, you created a web map including real-time and near-real-time layers, some with moving locations and others with updating attributes or photos. This web map provides timely information for real-time emergency management.

Next, you will create a dashboard based on this web map.

6.2 Create an operations dashboard

1. Open a web browser, go to ArcGIS Online (**https://www.arcgis.com**), and sign in. If you are continuing from the last section, click the Home drop-down menu and select Home.

2. In the page header, click the Apps button ⣿ , and choose Operations Dashboard.

You will be directed to the Operations Dashboard website.

3.   Click Create Dashboard.

4.   Enter **Redlands Emergency Dashboard** as the title, and **Redlands, Dashboard**, and **GTKWebGIS** as the tags, and then click Create Dashboard.

You are presented with an empty dashboard. Next, you will add a map to the dashboard.

5.   Click the + button and choose Map.

6.   Choose the Redlands Emergency Map you just created.

7.   Under Map Tools, select Legend, Layer Visibility, Basemap, and Zoom In/Out, and click Done.

Your web map displays in the dashboard. Next, you will add a gauge to display the count of the current open incidents.

8. Click the Add button, and select Gauge.

9. For layer, choose Reported Incidents – Dispatch.

10. Click the Data tab, and then click the + Filter button.

11. Set the filter as Operational Status > equal > Open as shown.

The default statistic option is Count, with 0 and 100 as the fixed minimum and maximum values.

12. Click the Gauge tab, change the Style as Meter, and leave the Shape as Horseshoe.

Data	**Gauge Options**	
Gauge	Style	Progress · **Meter**
General		
	Shape	**Horseshoe** · Half Donut

13. Click the + Guide button at the bottom three times, and define them as:

- From **0%** to **33%**, use the green color to indicate there aren't many open incidents.
- From **33%** to **67%**, use the orange color to warn that there are many open incidents.
- From **67%** to **100%**, use the red color to alert that there are too many open incidents.

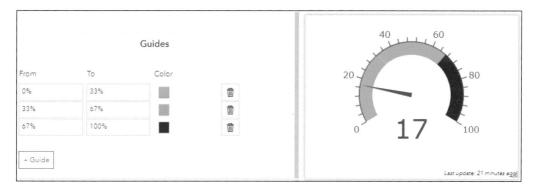

14. Click the General tab and set Title as **Open Incidents**.

Data	General Options	
Gauge	Name	Gauge (1)
General	Title	

B *I* U A▾ A▾ | ≡ ≡ ≡ ≡ | :≡ ≡ ≡ | ⊕ ⊗ ▣ ⊞

Normal ▾ | Default ▾ | *I*ₓ | Source

Open Incidents

15. Click Done to finish configuring the gauge.

A gauge element is added to the dashboard.

16. Click the Save button to save your dashboard.

In the next steps, you should save your dashboard regularly to avoid losing your work. Next, you will add a list element to list the 10 most recent incidents.

17. Click the + button and select List.

18. Select the Reported Incidents - Dispatch layer.

19. Click the Data tab, set Maximum Features Displayed as 10, and click + Sort.

Data	**Data Options**		Show data table
List			
	Using 'Reported Incidents - Dispatch' layer		Change
General			
	Filter	+ Filter	
	Maximum Features Displayed	10	
	Sort By	+ Sort	

20. For Sort By, choose the Open Date field, and click Descending.

	Open Date		🗑
Sort By	Ascending	Descending	
	+ Sort		

21. Click the List tab. In the Line Item Text tool bar, click { } to select the Incident field, hit Enter to move to the next line, and click { } again to select the Open Date field. Leave Line Item Icon as Symbol.

22.   Click the General tab and set Title as **10 Most Recent Incidents**.

23.   Click Done.

A list element displays on the dashboard.

Next, you will arrange the layout so that the list displays under the gauge.

24. Point to the blue bar in the upper-left corner of the list element to get its menu. Click and hold the Drag Item button, move the List element toward the bottom-center of the gauge element, and release your mouse to dock List element as a row.

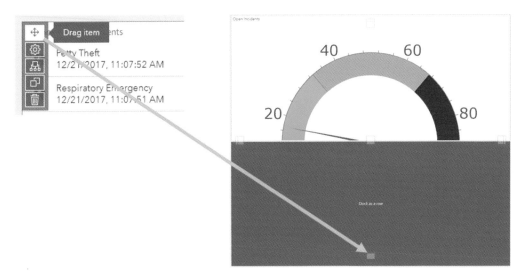

Next, you will resize the elements to have the map take up most of the screen.

25. Point to the vertical border on the left side of the map until your cursor changes to a crosshair, and then drag the border to the left until the map is your desired size.

26. Similarly, resize the gauge element so that the list gets more space.

Next, you will add a pie element to display the numbers of open incidents versus closed incidents.

27. Click the + button and select Pie Chart.

28. Select the Reported Incidents - Dispatch layer.

The Data tab and the Data options display by default.

29.   For Category Field, choose Operational Status. Leave Statistic as Count.

Categories From	Grouped Values	Features	Fields
Category Field	Operational Status		▽
Statistic	Count		▽
Sort By	+ Sort		

30.   Click the Chart tab, and set Labels Offset to **5** px.

31.   Click the General tab and set Title as **Open vs Closed Incidents**.

32.   Click Done.

A pie chart displays in the dashboard. Next, you will move the pie chart to the right side of the map.

33.   Point to the upper-left corner of the pie chart to get its menu. Click and hold the Drag Item button, move the pie chart element to the right side of the map, and dock it as a column.

Next, you will add a column chart to display the counts of every type of incident.

34.   Click the + button and select Serial Chart.

35.   Select the Reported Incidents - Dispatch layer.

The Data tab and the Data options display by default.

36.   For Category Field, choose Incident. Leave Statistic as Count.

37.   Click the Chart tab. For Orientation, select Horizontal.

38.   Click the General tab and set Title as **Incidents by Categories**.

39.   Click Done.

Next, you will move the column chart to the top of the pie chart.

40.   Point to the blue bar in the upper-left corner of the column chart to get its menu. Click and hold the Drag Item button, move the column chart element to the top part of the pie chart, and dock it as a row.

41.   Move the vertical board of the map to the right to increase the size of the map.

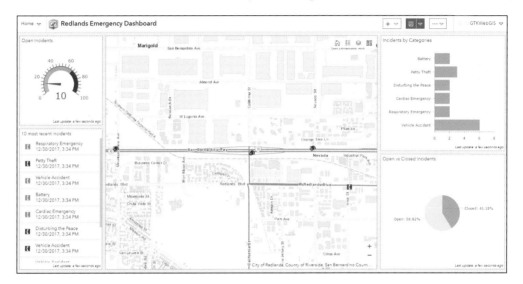

Next, you will add a header.

42.   Click the + button and choose Header.

43.   In the header appearance pane, set Size to Large, and set Background Color to light gray or any other color you like.

44.   Click Done.

6.3 Add actions and targets to your dashboard

When interacting with the dashboard, users can trigger such events as map extent change and selection change. These events can trigger actions (for example, filter), which can update the results in the target elements.

Next, you will add a selector to the header. The selector will allow users to filter the dashboard to show all incidents, only open incidents, or only closed ones.

1. Point to the blue bar in the upper-left corner of the header to get its menu and choose Add Category Selector.

The Selector tab and Selector Options display by default.

2. Under Selector Options, select the following settings:

- For Preferred Display Type, select Button bar.
- For Categories From, select Grouped Values, and select the Reported Incidents – Dispatch layer.
- For Category Field, select Operational Status.
- For None Option, select it.
- For Label for None, specify All.

As you choose the settings, you can preview the selector on the right side of the screen. Your selector should look like the illustration.

Next, you will specify the targets (in other words, to which elements the selections will be applied).

3. Click the Actions tab, click Add Action > Filter.

4. In the Actions pane, select the following settings:

    - Click Add Target and select the Reported Incidents – Dispatch layer.
    - Click Add Target again and select the List element.
    - Click Add Target again and select the Serial Chart.

Selector	**Actions**	
Actions	*When Selection Changes:*	
	Filter	Add Target ⌄
	╪ Reported Incidents - Dispatch	🗑
	≔ List (1)	🗑
	▊▊ Serial Chart (1)	🗑

5. Click Done.

The category selector displays in the header. Next, you will test the selector.

6. In the selector, click Closed, Open, and All to change the selections, and notice the filters apply to the map, the list, and the column chart.

7. Click Save to save your dashboard.

8. In the upper-left corner, click the Home button, and select Content.

9. In your content, find the dashboard you just created, and click it to go to its details page.

10. Click the Share button to share it with Everyone.

11. Click View Dashboard to explore your dashboard. Turn on the Traffic layer, and notice the moving vehicles, the emerging incidents, the traffic speed, and that the highway photos update as the underlying data changes.

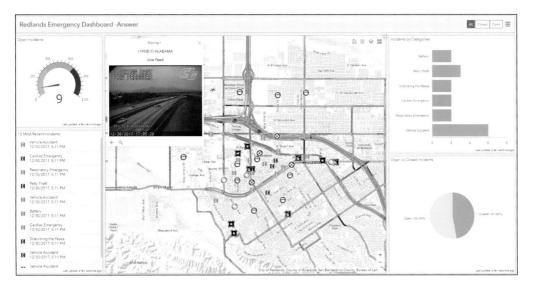

The URL of the current page is your dashboard URL that you can share with your audience. In this section, you created a dashboard that allows the city emergency management staff to obtain situational awareness and make sound decisions quickly by simply glancing at the app.

6.4 Create a time-enabled feature layer

A time-enabled feature layer requires a date field in the data. In this section, you will examine the date format of the lab data, publish a hosted feature layer, and enable time on this layer.

1. In a text editor or Microsoft Excel, open C:\EsriPress\GTKWebGIS\Chapter5\ US_Population_200_Years.csvl.

Rank	City	Population	Year	PopDate	Longitude	Latitude
1	New York, NY	33131	1790	06/01/1790	-73.91793106	40.70424544
2	Philadelphia, PA	44096	1790	06/01/1790	-75.1447934	39.99801526
3	Boston, MA	18320	1790	06/01/1790	-71.08912223	42.32160885
4	Charleston, SC	16359	1790	06/01/1790	-79.9862545	32.78930156
5	Baltimore, MD	13503	1790	06/01/1790	-76.61702283	39.30795807
7	Salem, MA	7921	1790	06/01/1790	-70.89850218	42.51685489
8	Newport, RI	6716	1790	06/01/1790	-71.31261926	41.48801068
9	Providence, RI	6380	1790	06/01/1790	-71.42214032	41.82355774
10	Marblehead, MA	5661	1790	06/01/1790	-70.86324017	42.49715389

The file lists the historic population of US major cities every decade from 1790 to 2000. The PopDate field has the population survey date for each record. To simplify the process, the dates for all years are the first day of June.

2. Study the data fields and rows. Notice each record has the spatial fields (in other words, Longitude and Latitude) and the temporal field (in other words, the PopDate field).

If a city existed at a specific time, the feature should have a geometry for that time. If the city also existed at another time, the city should have a geometry again in the data. This data requirement is clearly visible if the data is sorted by the city field, as illustrated. In this case, the same geometry gets repeated for each record of the same city. For features, such as vehicles and fire perimeters, which move, expand, or shrink, the same feature often has different geometries at different times.

Rank	City	Population	Year	PopDate	Longitude	Latitude
87	Akron, OH	42728	1900	6/1/1900	-81.51790247	41.0731642
81	Akron, OH	69067	1910	6/1/1910	-81.51790247	41.0731642
32	Akron, OH	208435	1920	6/1/1920	-81.51790247	41.0731642
35	Akron, OH	255040	1930	6/1/1930	-81.51790247	41.0731642
38	Akron, OH	244791	1940	6/1/1940	-81.51790247	41.0731642
39	Akron, OH	274605	1950	6/1/1950	-81.51790247	41.0731642
45	Akron, OH	290351	1960	6/1/1960	-81.51790247	41.0731642
52	Akron, OH	275425	1970	6/1/1970	-81.51790247	41.0731642
59	Akron, OH	237177	1980	6/1/1980	-81.51790247	41.0731642
71	Akron, OH	223019	1990	6/1/1990	-81.51790247	41.0731642
97	Akron, OH	217074	2000	6/1/2000	-81.51790247	41.0731642

Layers with spatial and temporal data for each record can be directly time-enabled in ArcGIS portals without any custom code. Such time-enabled layers can animate the time-series data in configurable apps directly.

3. Close Excel or the text editor. Do not save the CSV.

4. Open a web browser, go to ArcGIS Online (**https://www.arcgis.com**) or your Portal for ArcGIS, and sign in.

5. Click Content > My Content > Add Item > From my computer.

6. In the add an item from my computer window, make the following settings:
    - For File, browse to C:\EsriPress\GTKWebGIS\Chapter6\US_ Population_200_Years, and select it.
    - For Title, use the default, or specify a new one.
    - For Tags, specify **Historic Population**, **Major Cities**, and **GTKWebGIS**.
    - Make sure the check box next to Publish this file as a hosted layer remains selected.
    - Notice the PopDate field is recognized as a Date field.
    - For Time Zone, select UTC-05, which is US Eastern Time.
    - Click Add Item.

The time zone option for this decade-scale temporal data doesn't matter too much. For data at finer scales, it's more important to choose the correct time zone.

The item details page appears as your CSV file is being published as a hosted feature layer.

7.   On the item details page, look for the Layers section. Under the layer US_Population_200_Years, click Time Settings.

8.   In the Time Settings window, select the Enable time check box, select specific events in time, and select PopDate as the Time field. Click OK.

Time Settings   ✕

Enable time on this layer to visualize how the data changes over time using the time slider on the map.

☑ Enable time

The time data is recorded as:

⦿ specific events in time
Time field:

| PopDate | ▼ |

The latitude and longitude fields were needed for creating the feature layer, but they are not needed after your feature layer is created. The geometries of your points are managed by the feature layer itself internally. If you want to delete the latitude and longitude fields, click the Data tab, click the Longitude field header, and choose Delete. You can delete the Latitude field similarly.

9.   Click the Share button to share this layer with Everyone.

You have created a time-enabled feature layer. You will use this layer to animate the time-series data in the next section.

6.5 Animate time-series data in web maps and web apps

1.   Continuing from the last section, click Open in Map Viewer.

The layer is added to the map viewer. A time slider displays at the bottom of the map viewer. This indicates at least one of the layers in your web map is time enabled.

You are prompted with the Change Style pane. Next, you will style the cities based on their population.

2.   In the Change Style pane, select Population as the attribute to show, and click Options under Counts and Amounts (Size).

3.   For symbol size, set the Max to 96.

Increasing the max size can enhance visually the differences between cities with smaller populations and larger populations, as well as the differences when the population of the same city changes over time.

4.   Click OK and click Done to finish changing the layer style.

Next, you will configure the time animation playback speed, time span, time window, and slider labels.

5.   To the right of the slider, click the Configure button —|— .

6.   In the Time Settings window, select the following settings:

- Set Playback Speed to its fastest speed.
- Click Show advanced options.
- Under Time Display, set display data in 1 Decade intervals.
- Leave only display the data in the current time interval selected.
- Click OK.

Time Settings ⑦   ✕

Playback Speed

Slower   Faster

Time Span
Drag the slider handles or click a layer time line to set the Start and End time.

Layers   Layer Time Lines

US_Population_200_...

            Start Time: 5/31/1790   ▾   9:00 PM   ▾
            End Time: 5/31/2000   ▾   9:01 PM   ▾

Time Display
Specify the amount of data to display at one time.

Display data in 1   Decade   ▾ intervals.

As time passes   ⦿ only display the data in the current time interval.
                  ○ progressively display all the data.
Start playback at ⦿ start time.
                  ○ playback position saved with map.

                                              OK   CANCEL

7.   On the time slider, click the Play button. Try to understand the spatial and temporal patterns of US population change.

8. On the menu bar, click Save to save your web map.

- For title, specify **US population change from 1790 to 2000**.
- For Tags, specify **Historic Population**, **Major Cities**, and **GTKWebGIS**.
- Click Save Map.

Save Map   ✕

Title:	US population change from 1790 to 2000
Tags:	Historic Population ✕ Major Cities ✕ GTKWebGIS ✕
	Add tag(s)
Summary:	Description of the map.
Save in folder:	GTKWebGIS ▾

SAVE MAP   CANCEL

Next, you will create a web app from this web map.

9. On the map viewer menu bar, click the Share button ⬒.

10. In the Share window, share your web map with Everyone (public).

11. Click Create a Web App.

The ArcGIS configurable apps gallery displays. You will need to choose an app that supports time animation, such as the Time Aware app. You can also create your web app using ArcGIS Web AppBuilder with its Time widget. Here you will use the Time Aware web app template.

12.   In the configurable apps gallery, search for Time.

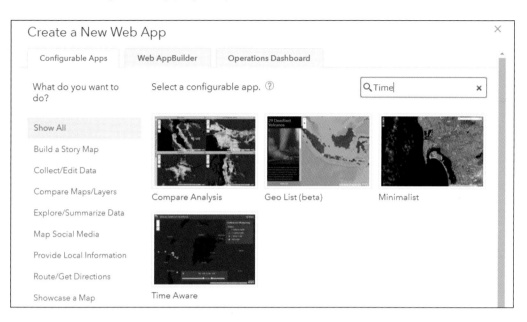

13.   Click the Time Aware template, and click Create Web App.

14.   In the Create New App window:

- Leave the default values for title and tags.
- Notice this app is shared with Everyone, same as the web map.
- Click Done.

You will be directed to the configuration mode of the app.

15.   Click the Time Settings tab, and select the following settings:

- Select Add tick marks to slider.
- Unselect Show end date.
- For Pre-defined date format, select year only (for example, 2015).

16.   Click Save.

17.   Click Launch.

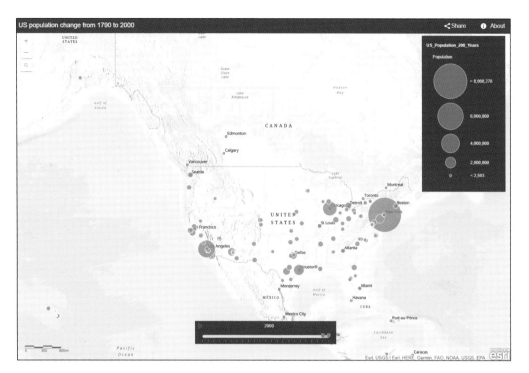

Your app displays in a new window. When you click the Play button on the time slider, the spatial and temporal patterns of US population change unfold. The US population mostly concentrated in the east in the early years and grew rapidly in the western states in the 1990s. While New York remains the largest city, Los Angeles emerged in the 1840s, grew rapidly in the 1990s, surpassed Chicago in the 1980s, and became the second largest city. The time animation effect revealed a spatial and temporal pattern that you can easily appreciate.

In sections 6.1 to 6.3, you created a dashboard app that displays continuous data feeds and photos in near real-time. Emergency response managers can obtain comprehensive views of a situation at a glance and make crucial decisions quickly. In sections 6.4 and 6.5, you created a time-enabled feature layer and a web app with time animation. The animation effect clearly exemplified the spatial and temporal patterns of US population change over 20 decades.

QUESTIONS AND ANSWERS

1. Can I publish a GeoEvent service to ArcGIS Online?

   Answer: No, but you can have your GeoEvent service publish outputs to ArcGIS Online.
   To publish a GeoEvent service, at a minimum you must have a base deployment of ArcGIS Enterprise with the server license as GeoEvent Server. Your GeoEvent service runs on your ArcGIS Enterprise. You can, however, configure your GeoEvent service to publish the output, for example, a feature set, to a hosted feature layer in ArcGIS Online. The output can then be used in web maps and web apps, and be made available to your end users in real-time.

2. How can I publish a stream service to push GeoEvent data?

   Answer: In GeoEvent Manager, create an output using the Send Features to a Stream Service connector, fill in the appropriate parameters, and publish the stream service. Then, create and publish a GeoEvent service that uses this output.

3. Instead of having a time stamp or a duration for each record, I have my temporal data in separate fields, such as Y1980, Y1981, Y1982, and so on. Can I create a web map to animate such data?

   Answer: To animate such data using the out-of-the-box web apps, you must reformat the data to switch the data stored in temporal fields or columns to rows. You can do so using the Transpose Fields tool.

4. In Operations Dashboard, can indicators and charts be configured to reflect only what is shown in the map extent? In other words, the indicators and charts would update as the user navigates the map.

   Answer: Yes. If you open the "configure actions" dialog for the map, you can choose what to filter when the map extent changes. Click Add Target, then just select the indicators and charts in the list of available target elements.

ASSIGNMENTS

Assignment 6A: Create a dashboard app to monitor the 311 incidents reported.

Data: Provided as a hosted feature layer. Search for "point incidents dashboard owner:GTKWebGIS" in ArcGIS Online.

Hints:

- Add the provided feature layer to a web map, and set the layer to refresh at an interval, for example, 0.2 minutes, and save your web map.
- Create a dashboard based on this web map.
- Use Collector for ArcGIS to open this web map and collect data, put your name, nickname, or initials in the attributes, and monitor the changes in your dashboard.

Requirements:

- Your dashboard should include at least a map, a gauge, a list, a column chart, a header, and a category selector.
- The category selector should at least filter the map, the gauge, and the list.

What to submit:

- The URL to your dashboard app

Assignment 6B: Create a web app to animate time-series data.

Data: Use the provided CSV file or your own data.

Hints:

- In ArcGIS Online, find the CSV file by searching for "helicopter, playback, owner:GTKWebGIS." Download it and examine the data format.
- Publish the CSV file as a feature layer and enable its time.
- Add this layer to the map viewer.
- Style the layer with the animating helicopter icon as you did in section 6.1.
- Play back the helicopter locations at a time interval of 3 seconds.
- Save your web map and create a web app with time animation.

What to submit:

- The URL to your web app

Resources

"ArcGIS GeoEvent Server: An Introduction," Josh Joyner and Sagar Ayare, https://www.youtube.com/
    watch?v=D4STBzp_iV8 (or http://bit.ly/2D2oNMp).

"ArcGIS GeoEvent Server help documents," http://server.arcgis.com/en/geoevent.

"ArcGIS GeoEvent Server: Leveraging Stream Services," https://www.esri.com/training/catalog/59b7fa8a
    17d9bf0910929d24/arcgis-geoevent-server:-leveraging-stream-services/ (or http://arcg.is/2BT5B3B).

"ArcGIS GeoEvent Server product introduction," https://www.esri.com/arcgis/products/geoevent-server.

"Chapter 9, Mapping the Internet of Things," *The ArcGIS Book*, Christian Harder and Clint Brown,
    http://learn.arcgis.com/en/arcgis-book/chapter9.

"GeoEvent community," https://community.esri.com/community/gis/enterprise-gis/geoevent.

"GeoEvent Server: Internet of Things (IoT)," Morakot Pilouk, Ming Zhao, and Josh Joyner, https://www
    .youtube.com/watch?v=n-UGR-QJ0m4 (or http://bit.ly/2AY4g9I).

"GeoEvent Server: Real-Time GIS," https://www.youtube.com/watch?v=KOTxytMJ42Q (or http://bit.ly/
    2mvC4FV).

"GeoEvent Server tutorials," http://enterprise.arcgis.com/en/geoevent/latest/get-started/geoevent
    -server-tutorials.htm.

"Introduction to GeoEvent Server," https://www.esri.com/training/catalog/5980e11fc42086479c7f3371/
    introduction-to-geoevent-server (or http://arcg.is/2DDPuaU).

"Making the Most of the Internet of Things: The Power of Location," https://assets.esri.com/content/dam/
    esrisites/media/pdf/14637-iot-ebook/G78166_Brand_IoT_ONLINE_dl_R5_WEB.pdf (or http://arcg.is/
    2CCsxUl).

"Monitor Real-Time Emergencies," http://learn.arcgis.com/en/projects/monitor-real-time-emergencies.

"Notifications in GeoEvent," https://www.esri.com/training/catalog/599313636a53461c0bc8f505/
    notifications-in-geoevent (or http://arcg.is/2DDITwX).

"Operations Dashboard for ArcGIS: An Introduction," Patrick Brennan, Chris Olsen, and David Nyenhuis,
    https://www.youtube.com/watch?v=paiEdhUaRAk (or http://bit.ly/2D3KlrT).

"Operations Dashboard for ArcGIS community," https://community.esri.com/community/gis/
    applications/operations-dashboard-for-arcgis.

"Operations Dashboard for ArcGIS help documents," http://doc.arcgis.com/en/operations-dashboard.

"Oversee Snowplows in Real Time," http://learn.arcgis.com/en/projects/oversee-snowplows-in-real-time.

"Real-Time GIS and GeoAnalytics," Jo Fraley, https://www.youtube.com/watch?v=OaodLd--Muk (or
    http://bit.ly/2DfrZr4).

"We Are Limited Only by Our Collective Imaginations—Jack Dangermond," Sarah Wray, http://www.esri
    .com/esri-news/arcwatch/0815/we-are-limited-only-by-our-collective-imaginations-jack-dangermond
    (or http://arcg.is/1Qhvk5S).

CHAPTER 7

3D web scenes

We live in a 3D world, so we typically find it easier to understand and analyze the world using 3D maps. 3D and its advanced forms, VR and AR, are making GIS more compelling, immersive, and intuitive. In ArcGIS, 3D web maps are referred to as *web scenes*. This chapter introduces the basic terminology of web scenes and the types of scene layers; discusses how to create scene layers using ArcGIS Pro and Drone2Map™ for ArcGIS®, and how to create web scenes using Scene Viewer; and then introduces 3D-related fields including VR, AR, and indoor 3D maps. The tutorial section first explores various scene layers in Scene Viewer, and then teaches how to use Scene Viewer to create web scenes and 3D web apps with thematic and realistic symbols. The "Questions and answers" section provides simple tutorials on how to use ArcGIS Pro to create scene layers.

Learning objectives

- *Understand the basic terminology of web scenes.*
- *Know the main types of scene layers.*
- *View web scenes using Scene Viewer.*
- *Create web scenes using Scene Viewer.*
- *Configure 3D symbols in Scene Viewer.*
- *Create 3D web apps from web scenes.*
- *Know the technologies involved to create scene layers.*

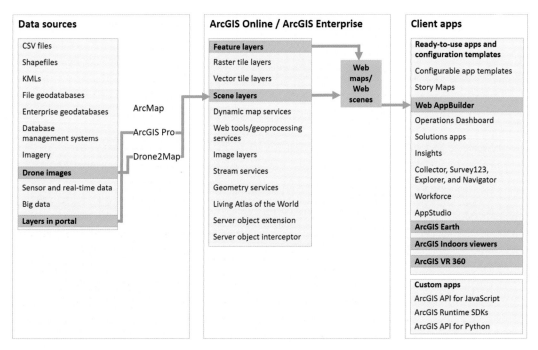

ArcGIS offers many ways to build web apps. The green lines in the figure highlight the technology presented in this chapter.

Basics of 3D GIS

In ArcGIS platform, 3D web maps are referred to as web scenes or 3D scenes. 3D brings an extra dimension into 2D maps, and thus often brings advantages in data visualization, analysis, and communication. Users often find 3D scenes interesting and more intuitive to interpret than 2D maps. These advantages give 3D GIS wider applicability in storytelling, urban planning, architectural design, defense simulation, filmmaking, and almost all industries. 3D GIS enables audiences to quickly understand the size and relative positions of objects. 3D GIS also enables designers to build flexible scenarios quickly and effectively to avoid costly mistakes in the building phase. Some 3D apps provide functions such as fly around and X-ray vision or radar vision, which allows users to see through buildings and the ground surface. In addition to visualization, 3D GIS can offer powerful analytical functions, such as visibility or viewshed analysis, sunlight and shadow analysis, and vertical zoning violation detection. Today, 3D GIS is a critical component for research frontiers and hot spots including digital cities, geodesign, indoor mapping, VR, and AR.

Just as a web map can include many layers, a web scene can also contain multiple layers, which can be feature layers, map image layers, image layers, raster tile layers, vector tile layers, and most importantly, scene layers.

Based on the visual effects, scenes can be grouped into two main types:

- **Photorealistic:** Aims to recreate reality by using photos to texture features. These types of scenes often use imagery as the texture and are extremely well suited for showing visible objects such as a city.
- **Cartographic:** Takes 2D thematic mapping techniques and moves them into 3D. These types of scenes often use attribute-driven symbols (extrusion height, size, color, and transparency) to display physical, abstract, or invisible features such as population density, earthquake magnitudes, flight paths, zoning laws, solar impact, and air corridor risks.

There are two view modes for scenes:

- **Global mode:** Displays features on a sphere. A global scene is best used for displaying phenomena that cover a large geographical area or wrap around the spherical surface of the earth.
- **Local mode:** Displays features on a planar surface. A local scene is best used for displaying or analyzing data at the local/city scale and underground.

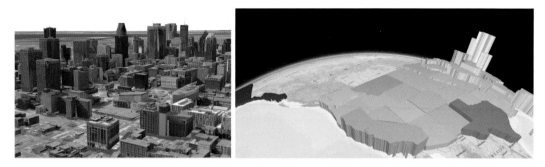

Left: A local and photorealistic scene showing the city of Montreal, Canada. Right: A global and cartographic scene displaying US states, with colors symbolizing the total area and extrusion heights symbolizing the population density.

Main elements in scenes

Scenes have four main types of elements:

- Surfaces: Surfaces are continuous measurements, typically elevation, with one value for a given x,y location. Surfaces provide the foundation for draping other content. Surfaces are often referred to as digital elevation model (DEM), digital terrain model (DTM), and digital surface model (DSM).
    - **DEM** is most of the time used as a generic term for DTMs and DSMs.
    - **DTM** represents the bare ground surface without any objects like plants and buildings.
    - **DSM** represents the ground surface including objects on it.

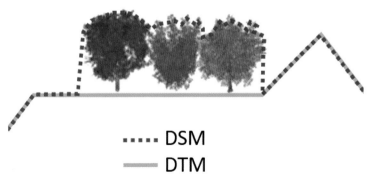

Differences between DTM and DSM.

- **Features:** Live on, above, or below the surfaces. These features are the operational layers of your 3D app.
- **Textures:** Provide exterior or interior covers of your 3D features. Textures often use aerial imageries or cartographic symbols.
- **Atmospheric effects:** Examples include lighting and fog.

3D across ArcGIS platform

ArcGIS platform provides a suite of 3D products to support the creation, visualization, analysis, and sharing of 3D scenes.

- **ArcGIS Pro:** A desktop app that provides comprehensive tools for managing 2D/3D data, authoring 2D/3D maps, and sharing 2D maps and 3D scenes.
- **Esri® CityEngine®:** A desktop app that provides advanced 3D creation capabilities. CityEngine can create photorealistic scenes manually. CityEngine can also create rule packages for generating large numbers of 3D objects in a batch. CityEngine has been used extensively in urban design and even filmmaking. Movies such as *Cars 2*, *Madagascar 3*, *Superman*, and *Zootopia* have used CityEngine to create massive urban environments.
- **ArcGIS Online and ArcGIS Enterprise:** Web GIS platforms that can host scene layers for online and on-premises 3D web GIS apps, offer Scene Viewer for creating and viewing web scenes, manage the access to web scenes and related layers, and provide Web AppBuilder and other ready-to-use apps for displaying web scenes.
- **ArcGIS® Earth:** A lightweight, easy-to-use 64-bit Windows desktop app for viewing and exploring 2D and 3D data, including KML, scene layers, and other layers.
- **ArcGIS API for JavaScript:** For developing browser-based custom 3D apps.
- ArcGIS Runtime SDK for iOS, Android, and Windows Phones: For developing native-based custom 3D apps.
- **ArcGIS 360 VR:** A mobile viewer that allows users to explore VR scenes generated by CityEngine.
- **Esri Labs AuGeo:** A mobile app that allows users to use ArcGIS data in an augmented reality environment.
- **ArcGIS Indoors and its 3D viewer:** A suite of tools for creating and managing indoor GIS data, layers, maps, and services. Its 3D viewer is a web-browser-based app that allows users to view indoor 3D scenes and perform indoor routing analysis.

Web scenes and web scene layers

A web scene can include 2D layers, such as feature layers, map images, vector tiles, raster tiles, and image services. While other 2D layers will drape on the surface, feature layers can be configured with 2D or 3D symbols.

A web scene can also include 3D layers, or scene layers. Scene layers, or scene services, are cached web layers that are optimized for displaying a large amount of 2D and 3D features. Scene layers follow the open Indexed 3D Scene Layer (I3S) format and provide a REST API to support client apps across all platforms. Scene layers include the following four types:

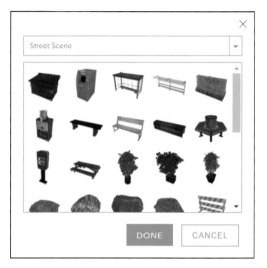

Feature layers can be configured to display in 3D symbols.

- **Point scene layers:** To ensure fast visualization in all clients, cached scene layers are used to display large amounts of point data not possible with a point feature layer. Once you add a point scene layer to your scene, you can configure it as you would any other point layers. Point scene layers are automatically thinned to improve performance and visibility at smaller scales. Automatic thinning means that not all features are displayed at small scales; as you zoom in, additional features are displayed. For example, you may use a point scene layer to display all the trees in a city. A point scene layer can be created using the Create Scene Layer Package geoprocessing tool, which generates an .slpk file on disk. This file can then be previewed directly inside ArcGIS Pro and then uploaded to Portal for ArcGIS or ArcGIS Online as a scene layer.
- **Point cloud scene layers:** These layers provide a fast display of large volumes of symbolized point cloud data, which are typically collected using Light Detection and Ranging (lidar) or generated using Drone2Map from drone images. Lidar, primarily used in airborne laser mapping applications, is emerging as a cost-effective alternative to traditional surveying techniques such as photogrammetry. Lidar is an optical remote-sensing technique that uses laser light to densely sample the surface of the earth, producing a cloud of points with highly accurate x,y,z measurements. Point cloud layers can be styled based on true color, class code (for example, ground, road, and water), elevation, and intensity values. A point cloud scene layer can be converted to a scene layer package and uploaded to Portal for ArcGIS or ArcGIS Online as a scene layer.

Left: A point layer displays in true color. Right: The same scene layer displays based on elevation.

- **3D object scene layers:** These layers can be used to represent and visualize 3D objects, such as textured or untextured buildings. 3D models can be created in ArcGIS Pro or Esri CityEngine, manually or automatically using procedural rules. Procedural modeling is more cost-effective than manual modeling, which is often labor intensive. Procedural rules can use 2D GIS data, including building footprints and attributes such as floors, roof types, and wall materials, to generate 3D objects for an entire city in a batch. To create a 3D object scene layer, you can convert a 3D object layer to a multipatch layer and then share it to Portal for ArcGIS, or further convert the multipatch layer into a scene layer package and then upload it to ArcGIS Online.
- **Procedural modeling = Geometry + Attributes + Rules.**

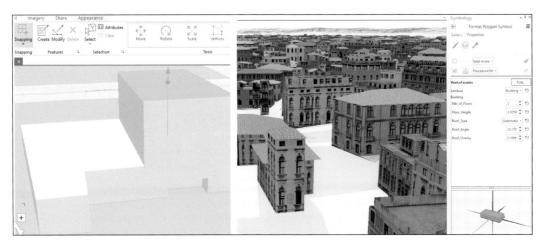

3D object layers can be created manually one by one (left) or in a batch automatically using procedural rules (right).

- **Integrated mesh scene layers:** Integrated mesh data is typically constructed from large sets of overlapping imagery, for example, using Drone2Map for ArcGIS. Drone2Map is a desktop app that turns raw still imagery from drones into valuable information products. Given a set of overlapping images, Drone2Map can identify the matching points on different images, stitch the images together based on these points, generate a point cloud, connect the points to construct triangle meshes, and apply realistic textures to create an integrated mesh layer. The mesh layer can be shared to ArcGIS Online or Portal for ArcGIS to produce an integrated mesh scene layer. In addition to textured mesh and point cloud layers, Drone2Map can produce other products such as DSM, DTM, and 2D orthomosaic layers. Drones can fly in areas that are difficult for humans to access because of factors such as size, terrain, and hazards. The information products that drones can produce are particularly useful for monitoring environmental changes and evaluating the impacts of natural disasters.

Drone2Map creates a textured mesh layer based on a set of drone images. The brown line represents the flight path of the drone. The blue dots represent the location where each image was taken. The texture mesh can be shared as an integrated mesh scene layer.

Create web scenes

ArcGIS Scene Viewer, ArcGIS Pro, and Esri CityEngine are the primary tools for creating web scenes. The first one is much easier to use, and the latter two provide more advanced authoring and geoprocessing capabilities in configuring 3D symbols. The following general steps are taken to create web scenes:

1. Choose global or local scene if using Scene Viewer and ArcGIS Pro.
2. Choose a basemap.
3. Add layers:
   - ArcGIS Scene Viewer can use web GIS layers, including scene layers, elevation layers, feature layers, map image layers, tile and vector tile layers, and tiled image service layers.
   - ArcGIS Pro can use web GIS layers and local data layers, including shapefiles, KMLs, layer packages, multipatch layers, and other layers in file or enterprise geodatabases.
4. Configure layers: such as layer styles, labels, and pop-ups.
5. Capture slides or bookmarks.
6. Save and share your scene.
   - For ArcGIS Scene Viewer, the scene is directly saved in ArcGIS Online or Portal for ArcGIS.
   - For ArcGIS Pro and CityEngine, the Share step involves publishing the web scene and its layers to ArcGIS Online or Portal for ArcGIS. Some layers need to be converted to multipatch type and then scene layer package type before sharing.

Choose global or local   **Switch basemap**   **Add layers**

Configure layers   **Capture slides**   **Save and share scene**

General steps to create web scenes.

Once web scenes are created, they can be viewed in many commercial-off-the-shelf (COTS) client apps and custom client apps.
- COTS clients include the following apps:
  - Scene Viewer.
  - Web AppBuilder for ArcGIS 3D mode.
  - ArcGIS configurable app templates in the Display a Scene (3D) category.
  - ArcGIS Earth: Compared to ArcGIS Scene Viewer, ArcGIS Earth supports KML and local data, including raster data, displays data on global scenes, and provides sketch and annotation capabilities.
  - ArcGIS Indoors 3D viewer.
- Custom client apps can be developed using ArcGIS API for JavaScript and ArcGIS Runtime SDKs.

A mobile app developed using ArcGIS Runtime SDK can display 3D web scenes, perform line-of-sight analysis (left) and perform viewshed analysis (right). Green lines and areas are visible and red are invisible. The radial red lines in the viewshed are areas blocked by the light poles.

VR and AR

VR and AR are intended to bring a location to the user in a more interesting way, allowing the user to interact and discover. Every day, more users and developers are embracing this new technology to create great applications. Understanding their uses and the business needs for them is important to create engaging and useful solutions.

VR

VR is a computer technology that uses headsets or multi-projected environments to generate 3D views, sounds, and other sensations that simulate a user's physical presence in a virtual or imaginary environment. A person using VR equipment can "look around" the artificial world, move around in it, and interact with virtual features or items.

VR provides a new and engaging way to visualize GIS data and interact with GIS data. It was a big advance for GIS maps to go from 2D to 3D. VR represents another leap from 3D maps. With 2D and 3D maps, users are still "outside" the maps. With VR, users are "inside" the maps. Used with a headset or helmet, VR can immerse users in the scenes generated from GIS data. Users not only see 3D scenes in front of them but also around them as they look around, turn, and walk. The VR environment is often live and interactive, allowing users to interact with the features in the view. The immersive feeling and interaction bring GIS data closer to users for better understanding.

ArcGIS 360 VR is a mobile-phone-based viewer that allows users to explore scenes generated by CityEngine in virtual reality. These scenes can portray different design alternatives for city blocks that can be compared or portray totally fictional scenes of historic or futuristic cities. These scenes can allow you to look underground, through walls, or even see thematic information overlaid on top of a representation of the real world. These scenes are published as VR experiences and stored as items on ArcGIS Online. A user can download the app to a phone, slide it into a GearVR headset, log in to ArcGIS Online, and load these VR experiences from the web.

Previously, VR users were required to purchase an expensive headset; now with the use of a mobile device and an inexpensive plastic or cardboard headset with plastic lenses, the user only needs to slide the mobile phone inside the headset with the screen toward the lenses. The application just needs to display two maps, one for each eye, with the correct lens distortion and separation from each eye.

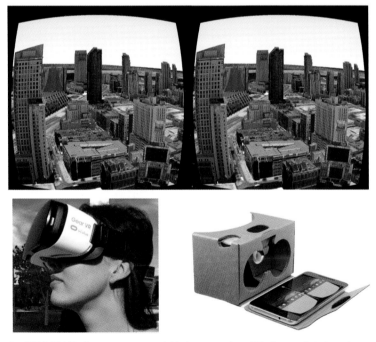

ArcGIS 360 VR allows users to quickly immerse into 3D city models by teleporting to viewpoints and comparing different urban design scenarios. The app is currently available for the Samsung Gear VR headset. Efforts have been made to make the app work with inexpensive VR cardboards.

AR

AR, also known as mixed reality, is a live view of a physical, real-world environment whose elements are "augmented" by computer-generated or extracted real-world sensory input such as sound, video, graphics, haptics, or GPS data. Even though AR and VR share some similarities, there is a major difference. AR enhances one's current perception of reality, whereas in contrast, VR replaces the real world with a simulated one.

Most of today's AR apps utilize a camera, GPS and other location services, compass, and gyroscope of smartphones. Smartphone vendors are beginning to provide AR development kits. These kits can detect planes like tables and floors using point clouds as anchors to add computer-generated objects into the camera image. Newer smartphones today can detect depth, producing good quality anchors to help position AR objects.

Coupled with GIS, AR apps can retrieve GIS data based on a mobile user's current location, looking direction, distance to a target, and then add the GIS data on top of the user's camera view. There are many cases when using AR to display geographic information is a perfect match. For example, AR apps can allow you to "see through" the ground (by overlaying underground pipelines over your camera view as you look down at the ground), "visit" a historical battlefield (by overlaying historic photos over your camera view as you look around in the field), and "see" the future (by overlaying a building's design over your camera view as you walk on the street).

Esri AuGeo is a mobile-phone-based viewer that augments reality by allowing users to overlay symbolized points over the view captured by their cameras. These points come from a feature layer hosted in ArcGIS Online. They could represent the location of fire hydrants in a 3D building, a label on top of each building in a neighborhood that has been searched, or the location of a water shutoff valve buried under 2 feet of snow.

Esri AuGeo displays facility information on top of the user's live camera view. The inset picture in the lower right corner is what users see on their phones.

Indoor 3D GIS

Moving GIS indoors—and integrating the indoors with the outdoors—holds enormous potential. Since its origin, the GIS domain has mostly been outdoors. The earliest applications of GIS were in forestry, resource management, and land-use planning, which are all about outdoors. GIS apps can help us find restaurants and hotels, but they offer almost no support for navigating in the complex indoor spaces of shopping centers and hospitals. In addition to indoor routing, GIS can also be helpful for managing indoor space utilization, indoor emergency response, guard deployment, and many other use cases. While the demand for indoor GIS is strong and clear, the development of indoor GIS faces two main challenges:

1.   GPS doesn't work indoors. There is a need for effective methods for determining indoor position.
2.   Indoor space is often in complex 3D formats. It's challenging to accurately model, store, share, and use indoor space data efficiently.

ArcGIS Indoors presents a set of tools for generating indoor GIS content, services, and applications. The core of ArcGIS Indoors is an ArcGIS Pro project that uses tasks to walk GIS professionals through the process of adding indoor floorplan information to the ArcGIS platform. Once the data is part of the ArcGIS system, it can be styled into 2D maps and 3D scenes, and scenes, published as services, consumed in a wide range of applications for indoor routing, facility management, security planning, and emergency responses.

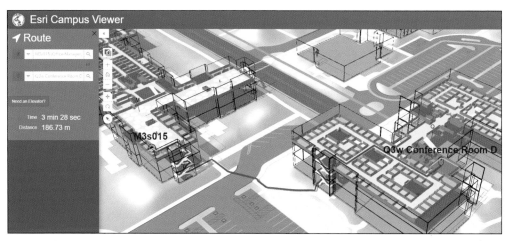

ArcGIS Indoors 3D viewer can display indoor 3D scenes and support 3D routing going through stairs and multiple buildings.

This tutorial

This tutorial teaches you how to create web scenes and 3D web apps.
- Section 7.1 uses ArcGIS Scene Viewer to explore various types of scene layers.
- Section 7.2 creates a web scene of typhoons, with the typhoon layer extruded based on wind speed.
- Section 7.3 creates a web scene of a park design, with a point layer styled with realistic 3D symbols.
- Section 7.4 illustrates how to enhance a web scene by editing the underlying 2D data.
- Section 7.5 introduces how to create web apps from web scenes.

Data: Provided via ArcGIS Online.

System requirements:
- ArcGIS Online for Organizations or ArcGIS Enterprise.
- A publisher or administrator account is required.

7.1   Explore web scenes using ArcGIS Scene Viewer

In this section, you will learn the basics of ArcGIS Scene Viewer and explore various types of scene layers.

1.   Open a web browser, go to ArcGIS Online (**https://www.arcgis.com**) or your Portal for ArcGIS, and sign in.

2.   Click Scene in the Main menu bar.

| Home | Gallery | Map | Scene | Groups | Content | Organization |

This step loads ArcGIS Scene Viewer.

3.   If prompted with a gallery of scenes, click the X button to close the window.

4. Click New Scene and choose New Global Scene.

5. Click the Basemap button ⊞⊞ to open the basemap gallery, and choose Imagery with Labels as the basemap for your scene.

Next, you will navigate the scene using the place/address search and using your mouse.

6. Click the Search button 🔍 to search for **grand canyon national park**, and press Enter.

> Search   ✕ | 🔍
>
>   grand canyon national park   ✕ | 🔍   ⊗

The scene will zoom to the Grand Canyon.

7. Click the X button to close the search box.

8. Move your cursor to the scene, note the Pan button ✛ is selected by default. If it is not selected, click this button to select it.

9. Explore and navigate your scene:

   • Click and hold the left mouse button to pan to a section of the Grand Canyon.
   • Use your mouse scroll wheel to zoom in. If you don't have a mouse wheel, you can click the + button in the navigation bar to zoom in.
   • Click and hold the right mouse button to rotate and tilt the scene.

You should be able to see 3D effects, similar to this illustration.

10.   Click the Compass button ⬆ to reorient your scene north.

11.   Click the Home button 🏠 to return to the initial camera position.

Next, you will explore different types of scene layers. You will find these layers by searching in ArcGIS Online.

12.   Under Layers click Add Layers.

13.   In the layer search box, type **Western Pacific Typhoons web scene**, and click Enter or the search button.

14. In the search result, click the import button ⓖ to import the found web scene into your scene.

This scene is a thematic type, with the heights of the cylinders representing wind speed and color representing wind pressure.

15. Move your cursor to the scene, notice the slides displayed on the bottom. Click a slide and note that the camera position changes and the layer visibilities change.

16. Under Add Layers, search for **Vancouver web scene**.

17. In the result, click the import button to import the found web scene.

18. Click the slides to navigate to different parts of the city. Notice this is an untextured 3D object scene layer.

19. Optionally, click the Remove button to remove this scene layer.

20. Under Add Layers, search for **Philadelphia web scene**.

21. In the result, click the Import button.

22. Click the slides to navigate to different parts of the city. Notice this is a 3D object scene layer with photo-realistic textures.

23. Optionally, click the Remove button to remove this scene layer.

24. In the Add Layers pane, search for **Iron Pagoda, Kaifeng, GTKWebGIS owner:webgis.book**.

25. In the result, import the found web scene. Notice this is an integrated mesh scene layer.

This layer was generated using Drone2Map for ArcGIS based on a series of drone photos.

26. Navigate close to examine the detail quality of the layer. Notice the top-down views are clear and sharp, but in tilted views, the sides of the Pagoda are not clear.

The photos used to generate the Iron Pagoda scene are nadir (straight down) images. Nadir images are best used to create 2D orthoimages, which are aerial photographs that show an area with consistent scale and minimal distortion. To produce high-quality 3D integrated mesh scene layers, oblique images should be used, the flight path should be around the target building, and the flight height must be appropriate to get clear images of the building sides.

27.   In the layer search box, search for **Milehigh scene owner:GTKWebGIS**.

28.   In the result, import the found web scene. Notice this is a point cloud scene layer.

29.   Click Done.

Next, you will change the style of the layer.

30.   In the Layers pane, click the point cloud layer.

31. For the drawing style, select True Color.

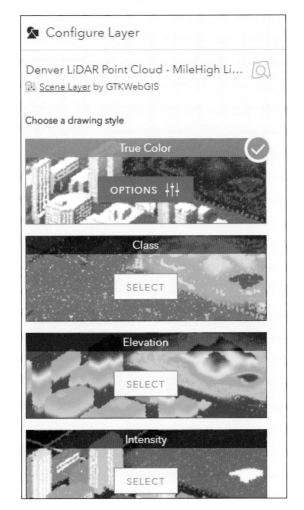

32. Zoom in and tilt the scene. Understand the points have elevations.

33. In the Configure Layer pane, select the Elevation style, and click the options button.

34. In the histogram, move the handles to stretch the color ramp, and notice the color changes according to the heights of the points.

35. Close the browser, and don't save your web scene.

7.2 Create a thematic web scene

1. Open a web browser, go to ArcGIS Online (**https://www.arcgis.com**) or your Portal for ArcGIS, and sign in.

2. Click Scene in the top menu bar.

3. If prompted with a gallery of scenes, click the X button to close the window.

4. Click New Scene and choose New Global Scene.

5. Click Add Layers.

6. Search for **Western Pacific Typhoons (2005) feature owner:GTKWebGIS**.

There is a feature layer found.

7. Click the Add button to add the found feature layer to your scene.

8. Click Done.

9. Navigate the scene to western Pacific and notice the typhoon paths that drape on the surface of the earth.

10. In the Layers pane, point to Typhoons Q1, click the Options button and select Configure Layer.

11. For the main attribute to visualize, choose wind_kph, which is wind speed in km per hour. For drawing style, choose 3D Counts & Amounts.

The typhoons display as standing cylinders in the scene.

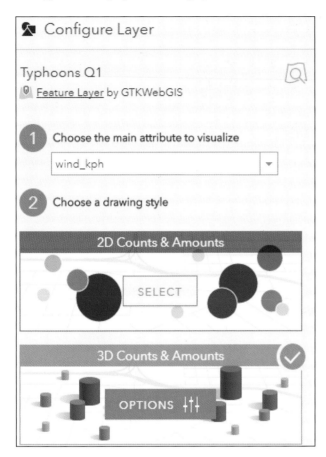

12.   Click Options.

13.   In the 3D Counts & Amounts pane, select the following settings:

- For Color, click the color ramp icon. In the color ramp gallery, scroll to select the yellow-red color ramp. Click Done.

- For Size, change the Min to 50,000, and max to 500,000.
- In the histogram, move the handles to select your desired color ramp and desired size ramp.
- Scroll down and turn on the Labels.
  - For Size, select large.
  - Click the Improve perspective toggle key to display labels more intuitively, with labels farther away having a smaller size.
- Click Done.

The typhoon labels display on top of the typhoons. If you rotate and tilt the scene, the labels remain upward.

14. Navigate the scene to a camera view position that you like.

This position will be the initial view position after the scene is saved.

15. Click Done.

16. Click Save Scene.

17. In the Save Scene window, select the following settings:

    - For Title, specify **Western Pacific typhoons Q1**.
    - For Tags, specify **Typhoons**, **3D**, and **GTKWebGIS**.
    - Click Save.

18. At the upper left corner of the page, click Home, and select Content.

19. In your content list, find the web scene you just created, select the scene, and share it with everyone, certain groups, or your organization.

7.3 Create a web scene of feature layers with 3D object symbols

1. Continuing from the last section, click New Scene, and choose New Local Scene.

2. If prompted, click Leave. This step will create a new blank local scene.

3. Click the Basemap button ⊞ , and choose Imagery with Labels as your basemap.

Next, you will add two feature layers.

4. In the Layers pane, click Add Layers.

5. In the layer search box, type **3d fun park features owner:GTKWebGIS**, and click Enter or the search button.

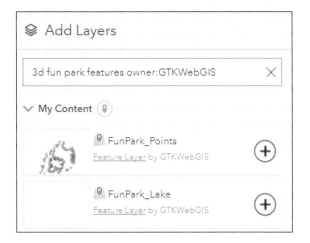

6. Click the Add buttons ⊕ to add these layers to Scene Viewer.

7. At the bottom of the Add Layers pane, click Done.

8. If the scene doesn't zoom to the new layers, click the Layers/Legend button, point to the FunPark_Points layer, and click the Zoom to button.

The scene zooms to the feature layers, and you will notice that the layers drape on the ground surface.

9. Click Save Scene.

10. Take these actions in the Save Scene window:

- Set the Title as **My Fun Park Design.**
- Set the Summary as **3D view of my fun park design.**
- Set the Tags as **3D**, **Fun Park**, and **GTKWebGIS.**
- Click Save.

Next, you will configure the style of the point layer based on the type, size, and rotation attribute fields of the layer.

11. In the Layers pane, point to the FunPark_Points layer, click the arrow ▼ , and click Configure Layer.

12. In the Configure Layer pane, for the main attribute to visualize, select ObjectType; for the drawing style, click 3D Types and then click the Options button.

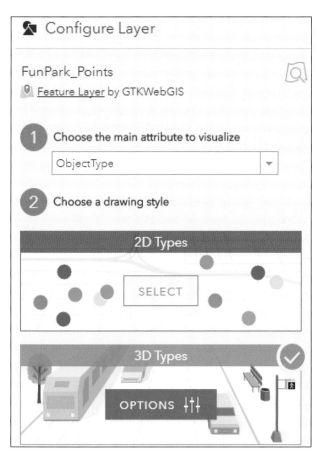

Next, you will specify the symbols for each type of object.

13. Under All symbols, for Size, select ObjectSize with ft as the unit.

14. Under All symbols, for Rotation, select the ObjectRotation field. Notice the rotation type is Geographic, which is measured clockwise from 12.

15. Tilt the scene and notice the points display in 3D symbols of various sizes.

Next, you will specify the symbol for each point type.

16.   Click the Palm category, and then the symbol box.

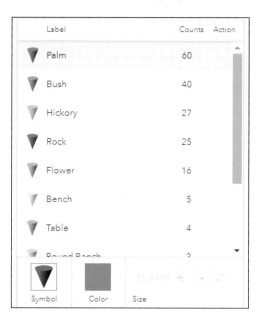

17.   In the symbol picker window, click the drop-down list and choose Vegetation.

The icons of the vegetation display in alphabetic order.

18.   Select the Banana Tree icon and click Done.

The points of the palm type now display in palm tree symbols in the scene.

19. Repeat the previous three steps to select the symbols for the rest of the point types.

   - For Bush, in the Thematic Vegetation group, select Amazon Sword Plant.
   - For Hickory, in the Vegetation group, select Red Hickory.
   - For Flower, in the vegetation group, select Orchid.
   - For Rock, in the Street Scene group, select Rock1.
   - For Bench, in the Street Scene group, select Park Bench2.
   - For Table, in the Street Scene group, select Picnic Table.
   - For Round Bench, in the Street Scene group, select Park Bench 4.
   - For Car, in the Transportation group, select Audi A6.
   - For Motorboat, in the Transportation group, select Motorboat.

20. Click Done, and Done again to finish changing the point layer.

21. Notice the points are all displayed in 3D symbols.

22. Click Save Scene and click the Save button.

23. Click Slides.

24. Navigate your scene to the picnic area, click Capture Slide, name it **Picnic area**, and click Done.

Next, you will style the lake layer.

25. Navigate your scene to the lake area.

26. In the Layers pane, point to the FunPark_Lake layer, click the arrow ▼, and click Configure Layer.

27. Click the Symbols arrow and select Change symbols.

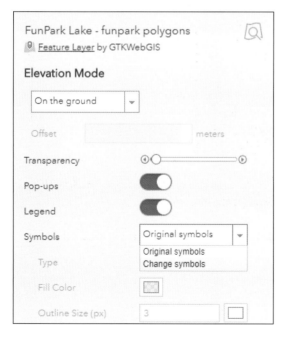

28. For Type, choose 2D Polygon. For Fill Color, choose Sky blue and set the transparency(%) to **0**. For Outline Size (px), set **7**. Click Done.

Symbols	Change symbols	▾
Type	2D Polygon	▾
Fill Color	☐	
Outline Size (px)	7	☐

29. Notice how the lake displays in blue and drapes on the ground.

30. Click the Basemap button and change the basemap to OpenStreetMap.

31. Click the Daylight button, choose the current date, and select Show shadows.

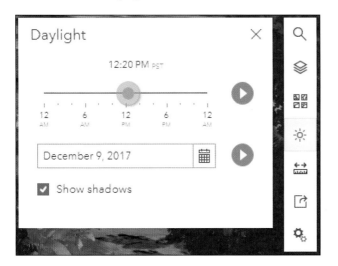

32. Click the time of day Play ▶ button to animate the sunlight as it cycles through the day. Notice the shadow pattern changes.

33. Click the Play button to pause, and set the time mark to be around noon.

34. Click Slides.

35. Click Capture Slide, name the slide **Lake**, and click Done.

36. Move your mouse to the scene, and in the slides list, click the Picnic area slide. Notice the changes in the view.

The camera view position changes to the picnic area, the basemap changes to imagery hybrid, and the shadows are not displayed. Slides in web scenes save the camera view position, basemap choice, shadows effects, and other settings such as layer visibilities.

37. Click Save Scene and click the Save button to save your scene.

38. Click the Home button in the upper left corner of the page and select Content.

39. Under My Content, find the web scene, and share it with Everyone, or certain groups, or only your organization.

40. Keep your browser open for the next section.

7.4 Enhance the web scene by editing 2D data and importing 3D object scene layers

A typical project will go through many iterations of design and evaluation. You are provided with the Fun Park points layer. In a real design project, you will need to edit your design, review the effects, and repeat the process many times.

The fun park scene you created in the last section is based on its underlying feature layers. In this section, you will make changes in the 2D points layer and review the effects in your 3D scene. You will also import a 3D object scene layer to your park.

1. In a new web browser, go to **http://arcg.is/2iLkAUO**.

This is a 2D web app pointing to the same feature layers used in your 3D scene.

2. Click the Open Attribute Table button ▰▰▰ .

3. Notice the FunPark_Points layer has three attribute fields: ObjectType, ObjectSize, and ObjectRotation.

Values of ObjectSize are in feet, and values in the ObjectRotation are in degrees, with 0 indicating no rotation, 90 indicating 90 degrees clockwise, and so on.

4. Click the Legend button to understand the point symbols.

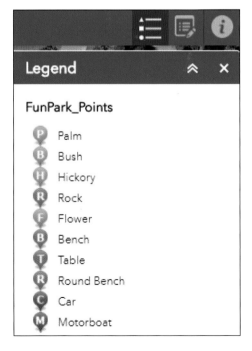

Legend

FunPark_Points

- Ⓟ Palm
- Ⓑ Bush
- Ⓗ Hickory
- Ⓡ Rock
- Ⓕ Flower
- Ⓑ Bench
- Ⓣ Table
- Ⓡ Round Bench
- Ⓒ Car
- Ⓜ Motorboat

Each point type is represented with the letter of the type.

5.   Click a few points on the map to review their types, sizes, and rotations in the pop-up. In the 3D web scene you created in the last section, review these points to understand how these attribute values control the 3D effects.

Next, you will edit the park design by editing the 2D data.

6.   Click the Edit button, and click a point type, for example, Motorboat.

7. Click on an appropriate location on the map, fill in the appropriate values for rotation and size.

8. Find the web browser showing the web scene you created in the last section and refresh the web page.

9. Navigate the scene to the place you added the point. Notice the point displays in a 3D symbol based on the attribute fields you specified.

10. In the 2D app, with the Edit widget open, click the point you just added. Change the size and rotation values, for example, to **20** and **175**.

11. In Scene Viewer, refresh the page, navigate to the point you just edited, and review the effects of your edits.

You may repeat these steps to add additional points and make edits to enhance your scene.

🖱 **Note:** To avoid the park points becoming too crowded, the FunPark_Points layer will be periodically restored to its original sample data.

Next, you will load a 3D object scene layer to this park.

12.   In Scene Viewer, under the Layers pane, click the Add Layers button.

13.   Search for **Fun park structures owner:GTKWebGIS**, click the Add button next to the matching item, and click Done.

14. Navigate the scene to examine the building, castles, and soccer field that were added to the scene.

The FunPark_Structures layer was generated from ArcGIS Pro using procedural rules and then published to ArcGIS Online as a scene layer. You can refer to the "Questions and answers" section for details.

Next, you will set the initial view of your scene.

15. Navigate the scene to a camera position that you would like users to see at the initial view.

16. Click Save Scene and click the Save button to save your scene.

7.5   Create a 3D web app using Web AppBuilder

1.   Open a web browser, go to ArcGIS Online (**https://www .arcgis.com**) or your Portal for ArcGIS, and sign in.

2.   Under Content > My Content, click Create > App > Using the Web AppBuilder.

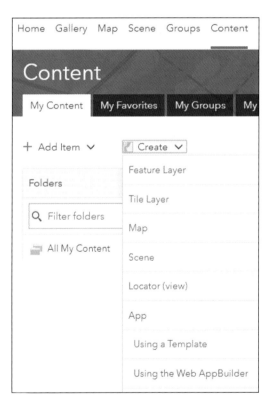

3.   In the Create a New Web App window, select the following settings:

    • Choose 3D.
    • For Title, specify **My Fun Park Design**.
    • For Tags, specify **3D**, **Fun Park, design**, and **GTKWebGIS**.
    • Click OK.

Note: Make sure you selected the 3D option.

4.   Under Theme, select the Box Theme, a color you like, and the second layout.

5.   Click the Scene tab, click the Choose web scene button, choose the fun park web scene you created, and click OK.

6.   If prompted with Map Switch Confirmation, click OK to continue.

7.   Click the Widget tab and click the Set the widgets in this controller link.

8.   Click the + button.

9.   In the Choose Widget window, select the Basemap Gallery, Layer List, and Legend widgets, and click OK.

10.   Click Save.

11.   On the preview side, click each button in your app to understand the capabilities of each button.

Future releases of Web AppBuilder will provide more 3D widgets with extended functionalities.

12.   In the upper left corner of the page, click Home > Content.

13.   Under My Content, select the app you just created, share it with Everyone, or certain groups, or your organization.

14.   Click your app, which leads you to the app details page.

15.   Click View Application.

This URL of your current page is your app URL. You can share this URL with your audience.

In this tutorial, you created web scenes by styling feature layers with 3D symbols. You edited your web scene by editing the underlying 2D data. This approach is simple and flexible, and works for a small amount of data.

The subject of 3D web scenes is too broad to cover in one chapter. This chapter introduced the basics of 3D web scenes. However, you should use ArcGIS Pro, CityEngine, and Drone2Map to create scene layers in the following cases:

- Your 3D projects have requirements for advanced symbols or exact building shapes with exact measures of roofs and windows.
- Your data size is huge.
- Your data is in point cloud format or drone image format.

For more information, refer to the "Questions and answers" and "Resources" sections.

QUESTIONS AND ANSWERS

1. I created photorealistic 3D models in CityEngine or other tools. How can I bring them into a web scene in ArcGIS Online or Portal for ArcGIS?

   Answer: You can export your 3D models to multipatch format, which is a GIS industry standard developed by Esri. This format uses collections of geometries to represent 3D objects. Multipatch features can be used to construct 3D features in ArcGIS, save existing data, and exchange data with other non-GIS 3D software packages such as Collaborative Design Activity (COLLADA) and SketchUp.

   You can add the multipatch layer into your scene in ArcGIS Pro by clicking the Map tab, Add Preset, and Realistic Building, and selecting your multipatch layer. Multipatch layers can be published to Portal for ArcGIS or converted to scene layer packages and then uploaded to ArcGIS Online. The resultant scene layers can be used in web scenes and 3D web apps.

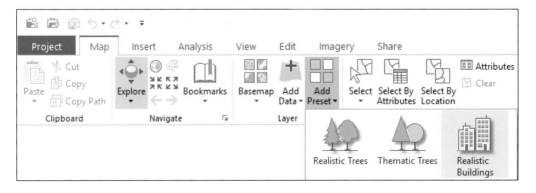

2. In ArcGIS Pro, how can I extrude a 2D line or polygon layer?

   Answer: Take the following steps:
   - In your scene, move the layer from the 2D group to the 3D layers group.
   - Click the Appearance tab on the top menu.
   - Remove the visibility range of the layer, if needed.
   - In the Extrusion group on the ribbon, choose an extrusion type, and specify the extrusion height. For example, set the extrusion height as population density times 100 meters.

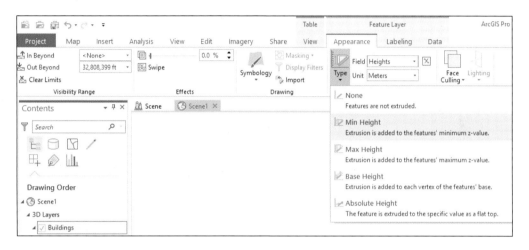

3. I like the soccer field, house, and castle imported into the fun park tutorial. How can I create those structures in ArcGIS Pro?

   Answer: The structures are created from a polygon feature layer using procedural rules. Here is a quick tutorial:
   - Create a scene project in ArcGIS Pro.
   - Add the C:\EsriPress\GTKWebGIS\Chapter7\FunParkData\Park.gdb\Structures layer to the scene.
   - In the Contents pane of your scene, move the layer from the 2D layers group to the 3D group.
   - In the Contents pane, right-click the Structures layer, and choose Properties.
   - In the layer properties window, set the elevation to be Relative to the ground using the baseElev field. Click OK.
   - In the Contents pane, right-click the Structures layer, and choose Symbology.
   - In the Symbology pane, choose Unique Values.
   - Click the symbol for SoccerField.
   - Click Properties and then the Layers button.
   - Click the polygon fill arrow and select Procedural fill.

- Click the Rules button and select C:\EsriPress\GTKWebGIS\Chapter7\ FunParkData\rpk\SoccerField.prk.
- Click Apply.
- In the Symbology pane, click back.
- Click the symbol for Castle and repeat the steps you did for the soccer field, but use Castle.rpk.
- In the Symbology pane, click back.
- Click the symbol for House, repeat the steps you did for the soccer field, but use VeniceFacades.rpk.

4. How can I publish my extruded line and polygon layer or a layer styled with procedural rules to Portal for ArcGIS or ArcGIS Online as scene layers?

    Answer: You can convert the layer to multipatch and then share the multipatch layer to Portal for ArcGIS. To publish to ArcGIS Online, you must further convert the multipatch layer to a scene layer package file, and then upload the scene layer package to ArcGIS Online. Here is a quick tutorial:
    - With your scene project open, click the Analysis tab and then click Tools.
    - In the Geoprocessing pane, search for Layer 3D to Feature Class.
    - Click the tool, use your layer as the input feature layer, and run the tool. The result is a multipatch layer, which can be shared to Portal for ArcGIS as a scene layer.
    - In the Geoprocessing pane, search for Create Scene Layer Package.
    - Click the tool, use the above multipatch layer as the input, specify an output file, and run the tool. The result is a scene layer package.

- Sign into ArcGIS Online, click Add Item > From my computer, select the previously mentioned scene layer package file, specify the title and tags, and click Add Item. The scene layer package will be uploaded to ArcGIS Online and published as a scene layer.

ASSIGNMENT

Assignment 7: Create a 3D web app of a fun community.

Requirements:
- Your subject can be a street, a residential area, a beach park, or others, but it must be in an area different from the fun park in the tutorial.
- Your web scene should have at least a point layer and a polygon layer.
- The 3D symbols of your point layer should be driven by attributes, such as ObjectType, ObjectSize, and ObjectRotation.
- Create at least two slides.
- Have some fun and make your scene appealing!

Tips:
- You can create feature layers in the many ways you have learned in the previous chapters. You can create your point feature layer by copying the FunPark_Points feature layer (The URL is stored in C:\EsriPress\GTKWebGIS\ Chapter7\FunParkData\FunPark_Points_URL.txt). You can also use ArcGIS for Developers (**https://developers.arcgis.com**, click the + sign, choose New Layer, specify the name, tags, geometry types, and attribute fields).

- Enable editing on the feature layers.
- Add the feature layers to a web map and add features to them.
- Add the feature layers to a web scene and style them.
- You may need to switch between the 2D editing window and 3D scene window many times to edit your data and to review the effects.

What to submit:
- The URL to your web scene
- The URL to your 3D web app

Resources

"ArcGIS Earth," http://www.esri.com/software/arcgis-earth.

"ArcGIS Indoors—Create a Campus Map and Scene," https://www.youtube.com/watch?v=2WWQmX5J2Mo (or http://bit.ly/2nMuquD).

"ArcGIS Online: Scene Basics," https://www.esri.com/videos/watch?videoid=TDjn13tP89o&channelid=UCgGDPs8cte-VLJbgpaK4GPw&title=arcgis-online:-scene-basics (or http://arcg.is/2kK9hjJ).

"Best Practices for 3D Scene Services," https://www.youtube.com/watch?v=ogVFF1PD4fQ.

"Chapter 6, Mapping the Third Dimension," *The ArcGIS Book*, Christian Harder and Clint Brown, http://learn.arcgis.com/en/arcgis-book/chapter6.

"CityEngine and ArcGIS 360 VR," https://www.youtube.com/watch?v=TSjQsHZT-q8 (or http://bit.ly/2A85gIB).

"Display a scene with realistic detail," https://learn.arcgis.com/en/projects/get-started-with-arcgis-pro/lessons/display-a-scene-with-realistic-detail.htm (or http://arcg.is/2A7MYHs).

"Drone2Map for ArcGIS," https://www.esri.com/products/drone2map.

"Drone2Map for ArcGIS: 3D Imagery Products," https://www.youtube.com/watch?v=Mxc_5oLzQWI (or http://bit.ly/2B3hsOP).

"Esri UC 2017: ArcGIS Online Scene Viewer—DYK?" https://www.youtube.com/watch?v=6Dpzk5W6eBg (or http://bit.ly/2Aqhukn).

"Esri UC 2017—Future of GIS and 3D." https://www.youtube.com/watch?v=jtg6PB1RBCU (or http://bit.ly/2kflRnr).

"Esri UC 2017: Walt Disney Animation Studios—Zootopia," https://www.youtube.com/watch?v=sOY3LY688QI (or http://bit.ly/2jmhas1).

"Getting Started with ArcGIS 360 VR," https://www.youtube.com/watch?v=fpsfS5vZlwM (or http://bit.ly/2jiVNv9).

"Get Started with Drone2Map for ArcGIS," https://learn.arcgis.com/en/projects/get-started-with-drone2map-for-arcgis (or http://arcg.is/2aMIF6Z).

"Get started with scenes," https://doc.arcgis.com/en/arcgis-online/get-started/get-started-with-scenes.htm (or http://arcg.is/2BoxHqd).

"GIS, BIM, and Indoor Mapping," https://www.youtube.com/watch?v=fdSHrkhbYXQ (or http://bit.ly/2kgpsRT).

"I3S specification," https://github.com/esri/i3s-spec.

"Looking Forward: Five Thoughts on the Future of GIS," Michael Goodchild, 2011, https://www.esri.com/news/arcwatch/0211/future-of-gis.html (or http://arcg.is/2AH38dZ).

"NJ Utility on Forefront with New Mixed Reality Application," Bill Meehan, https://www.esri.com/about/newsroom/publications/wherenext/nj-utility-on-forefront-with-new-mixed-reality-application (or http://arcg.is/2rXIG0G).

"Procedural symbology," http://pro.arcgis.com/en/pro-app/help/mapping/symbols-and-styles/procedural-symbol-layers.htm (or http://arcg.is/2nL73Sq).

"Sharing 3D Content Using Scene Layer Packages," https://www.esri.com/training/catalog/58471aa5fb83aeb761847d7f/sharing-3d-content-using-scene-layer-packages (or http://arcg.is/2AJy1wG).

"Sharing 3D Content with ArcGIS," http://training.esri.com/gateway/index.cfm?fa=catalog.webCourseDetail&courseid=2959 (or http://arcg.is/2Abwy0I).

"Share a web scene," http://pro.arcgis.com/en/pro-app/help/sharing/overview/share-a-web-scene.htm (or http://arcg.is/2yUP1AA).

"Scenes," http://pro.arcgis.com/en/pro-app/help/mapping/map-authoring/scenes.htm (or http://arcg.is/2vWJ0PZ).

"Using Drone2Map for ArcGIS," https://www.youtube.com/watch?v=XTB_bM10mSM (or http://bit.ly/2jBtiWv).

"View scenes in scene viewer," http://doc.arcgis.com/en/arcgis-online/get-started/view-scenes.htm (or http://arcg.is/2kQVzff).

"What is a scene layer?" https://pro.arcgis.com/en/pro-app/help/data/point-cloud-scene-layer/what-is-a-scene-layer-.htm (or http://arcg.is/2AI6jQY).

"Working with 3D Models in ArcGIS Pro," https://www.youtube.com/watch?v=XnhnV_dymsk (or http://bit.ly/2jDMaUK).

CHAPTER 8

Spatial analysis and geoprocessing

Web GIS is far more than mapping. As a key aspect of GIS, spatial analysis allows users to discover relationships, patterns, and trends in geospatial data. Traditionally, the power of spatial analysis was limited to GIS professionals with access to desktop GIS software, but Web GIS unlocked the power of spatial analysis for everyone. ArcGIS Online and ArcGIS Enterprise provide a rich collection of standard analysis tools. ArcGIS Enterprise provides tools specialized for big data analysis and allows you to share your own models and scripts as web tools to support tailored analyses. You can access these web tools through user-friendly interfaces in ArcGIS map viewer, Web AppBuilder for ArcGIS, Insights℠ for ArcGIS®, ArcGIS® API for Python, and other types of clients. These web-based analysis tools allow GIS professionals and nonprofessionals to gain geographic insights and make informed decisions. The tutorial in this chapter includes three case studies that will teach you how to use ArcGIS standard analysis tools, publish and use custom web tools, and perform big data analysis, respectively.

Learning objectives

- *Understand the web-based analysis tools provided in ArcGIS.*
- *Know the collection of variables available from ArcGIS Online.*
- *Understand the basics of Insights for ArcGIS.*
- *Create web apps that use ArcGIS Online analysis.*
- *Author and share web tools with ArcGIS Enterprise.*
- *Use web tools in web apps.*
- *Understand the workflow for big data analysis.*

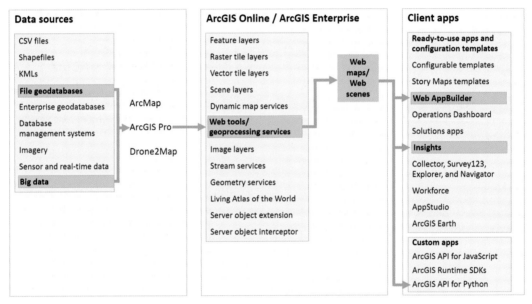

ArcGIS offers many ways to build web apps. The green lines in the figure highlight the technology presented in this chapter.

Spatial analysis and ArcGIS web tools overview

Spatial analytics holds great practical value for our personal life and enterprise operations. You most likely use spatial analysis every day, for example, by finding the optimal route to work or home using the apps on your smartphone. Banks, supermarkets, and real estate developers use spatial analytics to delineate market areas, estimate sales potential, and select facility locations. Your law enforcement agency uses spatial analytics to discover crime hot spots and decide where to deploy more police officers. The infinite applications of spatial analytics range from calculating the probable evacuation area after a hazardous chemical spill and predicting the track and strength of a gathering hurricane, to summarizing the land cover within a user-defined watershed and predicting a region's future development.

With the accelerating advances in sensor networks, IoT, and mobile and wearable technologies, humans are capturing huge amounts of digital data about the real world. These advances have led to big data, datasets that are so large or complex that traditional data-processing application software is inadequate to deal with them. Big data has four main characteristics in volume, variety, velocity, and veracity. These characteristics bring challenges to big data storage, transfer, visualization, analysis, and sharing.

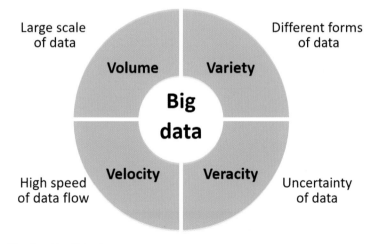

The four Vs of big data.

ArcGIS provides web tools for standard analysis, vector big data analysis, and raster data analysis. It also allows you to publish your own web tools or geoprocessing services. Refer to the next chapter for more details on raster analysis. This chapter focuses on the other analysis tools.

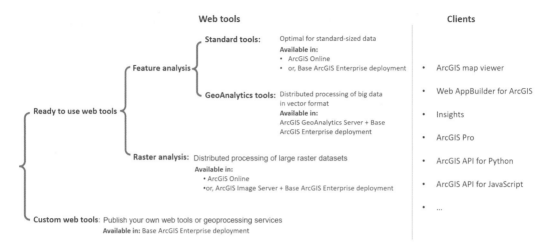

Web-based analysis tools provided by ArcGIS (left) and the clients that can use these tools (right)

ArcGIS Online and ArcGIS Enterprise differ and complement each other in the way they provide spatial analytics capabilities.

Compare ArcGIS Online and ArcGIS Enterprise analysis capabilities

	ArcGIS Online analysis tools	ArcGIS Enterprise	
		Ready-to-use tools	Custom web tools
Source of tools	Created and hosted by Esri. Ready to use.	Comes with software install. Hosted on your infrastructure. Ready to use.	To be created and hosted by organizations.
Difficulty level	Low. Easy to use.		Relatively high. Requires knowledge of ModelBuilder, Python or other scripts to create services.
Target users	Both nonprofessionals and professionals.		Professionals who provide web tools or services.
Data provided?	Yes (you can use your own data too).	No (you can use the data from ArcGIS Online).	
Cost credits	Yes.	No, unless you are referencing ArcGIS Online tools and premium content.	No. Costs organizations to purchase, install, and host.
System requirements	Publishers, administrators, and custom accounts with certain privileges.		Administrators and publishers (if enabled to publish geoprocessing tools).

ArcGIS web tools for standard analysis

ArcGIS Online and ArcGIS Enterprise provide similar standard analysis tools in six main categories.

Standard analysis **tools provided by ArcGIS Online and ArcGIS Enterprise**

Summarize Data		Find Locations	
	Aggregate Points		Find Existing Locations
	Join Features		Derive New Locations
	Summarize Nearby		Find Similar Locations
	Summarize Within		Choose Best Facilities *
Analyze Patterns			Create Viewshed *
	Calculate Density		Create Watersheds *
	Find Hot Spots		Trace Downstream *
	Find Outliers	**Data Enrichment**	
	Interpolate Points		Enrich Layer *

Standard analysis tools provided by ArcGIS Online and ArcGIS Enterprise (continued)

Manage Data		Use Proximity	
	Extract Data		Create Buffers
	Dissolve Boundaries		Create Drive-Time Areas *
	Merge Layers		Find Nearest *
	Overlay Layers		Plan Routes *
			Connect Origins to Destinations *

Note: Most of these tools are available in both ArcGIS Online and Portal for ArcGIS. Tools marked with "*" are not available in Portal for ArcGIS by default. The Portal for ArcGIS administrator needs to configure corresponding utility services (for example, using those provided by ArcGIS Online) to enable these tools.

Some of the standard ArcGIS web tools require underlying data provided by ArcGIS Online. For example, your analysis can calculate driving directions based on the street network, historic traffic, or real-time traffic data in the Living Atlas. You can also enrich (or add to) your location of interest with the location's population, income, housing, consumer behavior, tapestry, and thousands of other variable values found in the Living Atlas. Collecting, converting, and hosting such live and historic traffic variables and commercial variables normally requires a lot of time and money. By providing the data to you, ArcGIS Online makes such analysis much easier, cheaper, and more cost-efficient for many situations. Portal for ArcGIS can be configured to use such analysis capabilities and data from ArcGIS Online.

Workflow to use standard analysis tools

There are four general steps to use ArcGIS standard analysis tools. If you only use data from the Living Atlas for your analysis, you can skip the first two steps and start with the third step in the following list:

1. **Prepare data:** You may create your own layers or discover and use other users' layers. These layers can be vector data in many formats.
2. **Add to map:** Add your layers or the layers you discovered into the map viewer. You can create data dynamically by using map notes layers and then use the newly created data for analysis. If you are going to use layers from the Living Atlas, you do not need to add them to your map.
3. **Perform analysis:** You will determine the appropriate tool(s), specify the appropriate parameters, and run the tool(s).
4. **Review and interpret results:** The results are often in the format of hosted feature layers or related tables. The results are automatically added to the web map with pop-ups configured. You can review and accept the results or adjust your parameters and run the tools again. As a part of your web map and app, the results can be shared with your web audience.

There are four general steps to use ArcGIS Online and ArcGIS Enterprise standard analysis tools.

ArcGIS web tools for big data analysis

Big data is too big for a single desktop machine to process and too big to be transferred over the network efficiently. Legacy system architecture is not adequate to meet the challenges of big data. ArcGIS® GeoAnalytics Server harnesses the computing power of multiple machines and distributes big data analysis across multiple machines; this can process massive quantities of data quickly. GeoAnalytics Server must be deployed in addition to Base ArcGIS Enterprise deployment. At least one ArcGIS Data Store must be configured as the spatiotemporal type. As your data size grows, you can deploy more GeoAnalytics Servers and data stores onto more servers to scale out your system.

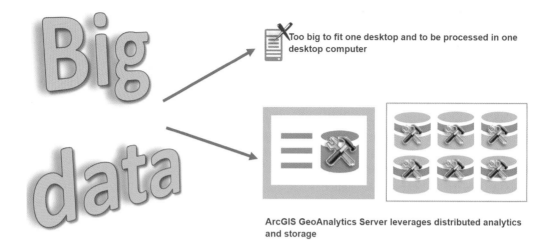

Too big to fit one desktop and to be processed in one desktop computer

ArcGIS GeoAnalytics Server leverages distributed analytics and storage

Big data is too big to be loaded in the memory of one desktop or processed by one computer (top). ArcGIS GeoAnalytics Server uses distributed computing to analyze big data (bottom).

Once deployed, GeoAnalytics Server provides a series of out-of-the-box and ready-to-use tools. These tools are arranged in categories including summarize data, find locations, analyze patterns, use proximity, and manage data. These tools emphasize spatiotemporal analysis, summarization, and aggregation to help users obtain intelligence from noises.

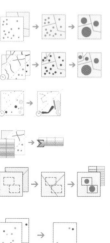

Summarize Data

Aggregate Points

Join Features

Reconstruct Tracks

Summarize Attributes

Summarize Within

Find Locations

Find Similar Locations

Analyze Patterns

Calculate Density

Create Space Time Cube

Find Hot Spots

Use Proximity

Create Buffers

Manage Data

Copy to Data Store

GeoAnalytics Server provides ready-to-use tools for big data analysis.

Workflow to use big data analysis tools

You'll follow five general steps when you use GeoAnalytics Server tools for big data analysis.

1. **Prepare data:** Your data sources can be the following sources:
   - File share: A directory of datasets on a local disk or network share
   - Hadoop Distributed File System (HDFS)
   - Hive: Metastore databases
   - Cloud store: An Amazon Web Services (AWS) Simple Storage Service (S3) bucket or Microsoft Azure Blob container containing a directory of datasets
2. **Register big data file share:** This step is typically done in ArcGIS Server Manager. This step serves two purposes:
   - Tells the GeoAnalytics Server where to retrieve the data.
   - Automatically generates a manifest file that identifies the geometry fields, time fields, and attribute fields.
3. **Edit big data file share if needed:** If the manifest file generated is not correct, you can edit the file to assign the correct fields for geometry and time.
4. **Run analysis:** You can select a tool, browse to a dataset in a big data file share registered, and run the tool.
5. **Review and interpret results:** The results are often in the format of hosted feature layers or related tables. The results are automatically added to the web map with pop-ups configured for you to review and share.

As you can see from these steps, you won't add big data to web maps directly as you would do when using the standard analysis tools. Big data is too large to be displayed on maps, and the maps would be too crowded and busy to make sense. However, the analysis results are much smaller and can be displayed and reviewed in web maps.

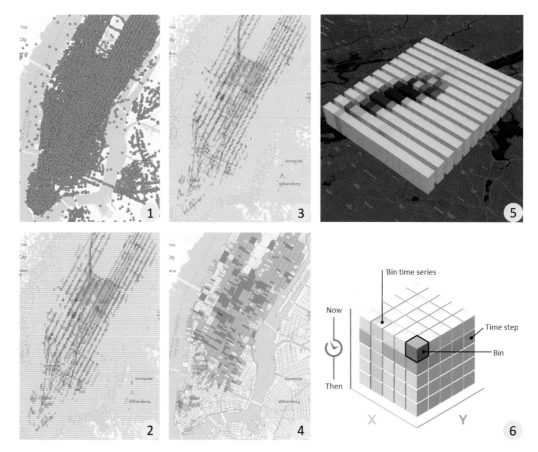

New York City had over 170 million taxi pickup locations recorded in 2010. The data size is 50G. A map of the raw data (1) took days to draw and the map is too crowded to display any patterns. Aggregation by squares (2), by hexagons (3), and by block groups (4) reveal the spatial patterns clearly. The space time cube (5 and 6) generated by aggregating these points reveals both the spatial and temporal patterns.

Custom web tools and geoprocessing services

The analysis tools provided by ArcGIS Online and ArcGIS Enterprise are essentially supported by geoprocessing services. In Portal for ArcGIS, a web tool is an item type, and behind the item is a geoprocessing service running in ArcGIS for Server. You can use ArcGIS Desktop, ArcGIS Pro, or

ArcMap, to publish your desktop tools as web tools in ArcGIS Enterprise. ArcMap allows publishers to publish tools as geoprocessing services to ArcGIS for Server. ArcGIS Pro allows publishers to publish a geoprocessing service to ArcGIS for Server, and at the same time create a web tool item in Portal for ArcGIS.

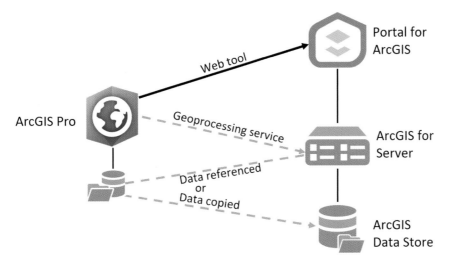

When sharing a tool from ArcGIS Pro to ArcGIS Enterprise, the data is copied to the server or referenced by the server. Then, a geoprocessing service is created in ArcGIS for Server and registered in Portal for ArcGIS as a web tool.

Typically, a web tool can include one or more tasks. Each task takes inputs, processes them, and returns meaningful and useful output(s) in the form of features, maps, reports, and files. If you have used local geoprocessing tools that run on your local desktops, you can think of geoprocessing services as tools running on server computers.

Steps to creating web tools

Creating a web tool requires you to author, run, and publish the execution plan of your tool.

Three steps to creating a geoprocessing service or web tool.

1. **Author your tool:** You typically create and document a geoprocessing tool using ModelBuilder or Python. In this step, you need to implement the workflow logic, specify the location of the required data, and define the input and output parameters. The output parameters define what web clients can receive and what your end users can access.
2. **Run your tool:** Once you've authored a tool, you must run it, and the tool must complete successfully.

3. **Publish the execution plan:** When the tool is finished running, it can be shared; sharing creates the web tool item in the portal and the back-end service on an ArcGIS Server federated with the portal. For publishing web tools, a key parameter is the execution mode. Execution mode defines how the web client interacts with the server and gets results from the executed task. This mode can be asynchronous or synchronous. Both forms of this mode are supported by Map Viewer, Web AppBuilder for ArcGIS, ArcGIS® API for JavaScript™, and other clients.

- **Synchronous:** Typically, this mode is for those tasks that execute quickly—in five seconds or less. If you set your service as synchronous, the client will wait for the task to finish to get the result. If a synchronous task takes too long to finish, the web client will get a time-out error.
- **Asynchronous:** This mode is typically for those tasks that take a longer time to execute. With this mode, the client must periodically ask the server if the task has finished and get the result from the server when the task has finished. This mode is recommended for most situations.

Python and ArcPy

You can use scripting languages such as Python to create your tools. Python is a free, powerful, cross-platform, and open-source scripting language included in a typical ArcGIS® Desktop and ArcGIS Server installation. Python often is used to automate workflows so you do not have to perform them manually. You can write Python script using any text editor (such as Notepad) or more sophisticated development environments. You can run a Python script inside ArcGIS or outside ArcGIS as a stand-alone app.

ArcGIS extends Python by providing ArcPy, a module that facilitates data analysis, data conversion, data management, and map automation. This module offers users ease and convenience in using Python by including such features as code completion (in which, for example, users can type a keyword plus a dot to get a pop-up list of properties and methods supported by that keyword) and reference documentation for each function, statement, module, and class.

ModelBuilder

In addition to Python, you can use ModelBuilder to create tools. ModelBuilder comes with ArcGIS Desktop, which you can use to create, edit, and manage models. Think of ModelBuilder as a visual programming language. Creating a model requires stringing together sequences of geoprocessing tools and connecting them with inputs and outputs.

Access to ArcGIS web tools

The ArcGIS web tools, including standard tools, GeoAnalytics tools, and custom tools, can be accessed in the portal map viewer, Web AppBuilder, Insights, Python API, JavaScript API, ArcGIS Pro, and other clients that use ArcGIS REST API. You will learn how the first two clients access ArcGIS web tools in the tutorial of this chapter and learn JavaScript API in the JavaScript chapter. The following several sections will introduce Insights and Python API.

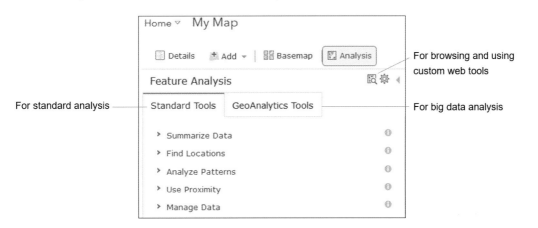

In Portal for ArcGIS map viewer, you can access the standard tools, big data analysis tools, and your custom web tools.

Perform analyses in Insights

Insights is an app that allows you to perform iterative and exploratory data analysis. You can answer questions with data from ArcGIS, Excel spreadsheets, and business databases by simply dragging the data into the app interface and performing the analysis. Insights integrates business intelligence-(BI)-like capabilities and user experiences to provide the ease-of-use for analyzing spatial and tabular data.

Insights bridges GIS analysis and business intelligence analysis.

Insights has the following key features:
- The user experience is drag-and-drop driven and intuitive.
- You can simultaneously visualize and analyze data in maps, summary tables, and an assortment of charts including bubble, histogram, scatter, time-series graph, and treemap, as well as traditional bar, line, and donut charts.
- Analysis and visualization happen at the same time on the same screen, thereby creating an evolved dashboard that stimulates questions and rapid iterations of analysis.
- Linked cards allow you to interact with more than one map or chart at a time (for example, click a bar on a chart and see related features light up on a linked card).
- Data from multiple sources can be blended and used in your analysis.
- Demographic data from ArcGIS will enrich your analysis.

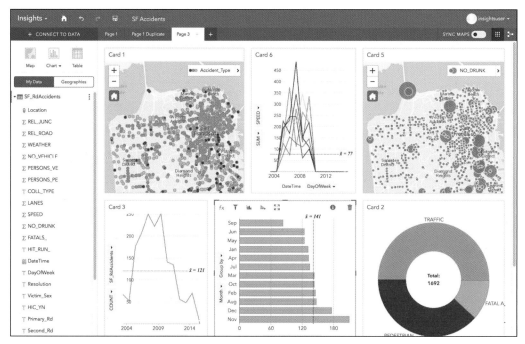

Insights enables simple drag-and-drop data exploration and analysis and visualization capabilities.

Perform analysis in Python API

Python API is a Python library for working with maps and geospatial data, powered by Web GIS. The API can be used in a browser-based interactive scripting environment called Jupyter Notebook. The notebook environment provides an interface to execute code, visualize portal items, users, and groups, as well as view web layers, maps, and scenes interactively. The API provides a simple and efficient way to access ArcGIS web tools to perform spatial analysis.

If you want to learn how to use web tools in ArcGIS API for Python, you can do so in a sandbox (go to **http://arcg.is/2zYJQBu**, click Try it live, and click New to create a new notebook), or install the API on your computer (follow the instructions at **http://arcg.is/2zYJQBu**, run the command jupyter notebook, go to **http://localhost:8888/tree**, and click New) and follow the samples at **http://arcg.is/2z2x3hi**.

In addition to using the analysis tools provided by ArcGIS, Python API, along with the Jupyter Dashboard, integrate well with the rich scientific Python ecosystem, including deep learning and artificial intelligence libraries, and can extend the analysis capabilities provided by ArcGIS.

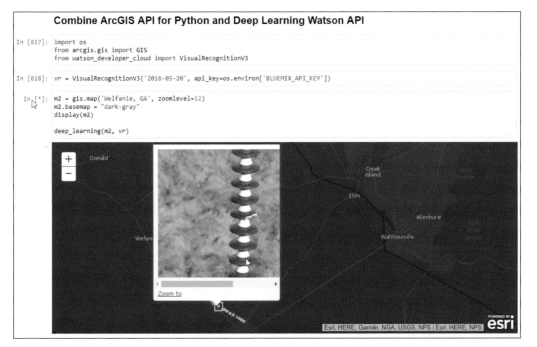

Combined with IBM Watson deep learning API, Python API can visually filter the hundreds of thousands of drone images taken along Georgia Power's 17,000 miles of transmission lines, automatically identify the broken, contaminated, or flashed insulators, and highlight the locations of the broken insulators on the map.

This tutorial

This tutorial includes three case studies. Instructors can choose one or two case studies based on the class time available and the software available. For example, you may turn the second and third studies into reading assignments if your school doesn't have ArcGIS Enterprise set up.

Case Study 1 (sections 8.1–8.2): Create a web app for selecting restaurant locations using ArcGIS Online analysis capabilities.

A company wants to open a new high-end, full-service restaurant. Company management must choose one of two possible recommended locations.

Functional requirements: The web app must have the following capabilities:
- Calculate the service area of each candidate.
- Obtain demographic information, including population, disposable income, sales potential, and leakage/surplus of restaurant service for each service area.
- Compare the demographics of the two recommended locations and help the company choose the location.
- Obtain the block group IDs of the service area in the selected location for a direct mail marketing campaign.

Data:
- Addresses for the two candidate locations.
- All other data will be derived from the Living Atlas.

System requirements:
- An ArcGIS Online administrator- or publisher-level account (with the privileges to run analysis and create hosted layers)

Case Study 2 (sections 8.3–8.6): Create a web tool and a web app using ArcGIS Enterprise for selecting factory locations. This case study is optional.

A company wants to build a factory in the US state of Alabama. You are contracted to build a Web GIS app that helps company executives select possible factory sites.

Site selection criteria:
- The new factory should be close to desirable areas determined by company executives.
- This factory will use a lot of water and should be built close to rivers.
- Factory products will require railroad transportation, so the factory should be close to railroads.

Functional requirements: Your web app should allow company executives to specify the following parameters:
- A point to indicate a location of interest.
- A distance that delimits the factory's distance from the location of interest.
- A distance that defines how close the factory should be to rivers.
- A distance that defines how close the factory should be to railroads.

Data:
- A file geodatabase containing the following two feature classes:
  - Main rivers in Alabama
  - Selected railroads in Alabama
- A map document named **Site_Selection.mxd** to display river and railroad layers.
- A toolbox named **Planning.tbx**, which contains a **Select_Sites** tool.

System requirements:
- ArcGIS Pro 2.1 or newer for designing your GP model.
- ArcGIS Enterprise 10.6 or newer for publishing and hosting your geoprocessing service.
- An ArcGIS Enterprise administrator or publisher account that can publish web tools and performance analysis.

Case Study 3 (section 8.7): Analyze the big data of New York City taxi cab drop-off and pick-up locations, and determine the spatial and temporal patterns of the drop-off locations. This case study is optional.

System requirements:
- ArcGIS Enterprise 10.6 or newer with ArcGIS GeoAnalytics Server role installed.
- An ArcGIS Enterprise administrator account for configuring the big data file share.

8.1 Create a web app using the Web AppBuilder Analysis widget

You have learned how to use Web AppBuilder in earlier chapters. Therefore, this section will be brief.

1. Sign in to ArcGIS Online (**http://www.arcgis.com**) or your Portal for ArcGIS.

2. Click Map to go to the map viewer.

3. On the toolbar, click the Basemap button, and choose the Streets basemap.

The Streets basemap emphasizes transportation features and easily allows you to see the service areas that must be calculated.

4. On the toolbar, click the Add button , and choose Browse Living Atlas Layers.

5. In the Living Atlas pane, search for tapestry, add the 2017 USA Tapestry Segmentation layer to your map.

The tapestry layer helps you determine the location of potential customers. In an actual project, you could add more layers that are relevant to your project requirements.

6. In the Contents pane, turn off the tapestry layer.

You will turn the layer on as needed in your web app.

7. Save your map as **Map for restaurant location selection**, and set the keyword tags as **tapestry** and **GTKWebGIS**.

8. On the toolbar, click the Share button, select Everyone, and click Create a Web App.

9. Click the Web AppBuilder tab.

10. In the Create a New Web App window, set the title as **Restaurant Location Selection**, and click Get Started.

11. Under the Theme tab, choose Box Theme, and select a layout.

You can select a different theme, but the theme must have the Attribute Table widget. You will need this widget in the analysis steps later.

12.   Click the Widget tab.

13.   Enable the Attribute Table widget by pointing to the widget and clicking the Eye button.

14.   Point to the Attribute table widget and click the pencil icon to configure it.

15.   In the Configure Attribute Table widget, select Allow exporting to CSV, and click OK.

16.   Click Set the widgets in this controller link.

You will see the **Legend** and **Layer List** widgets in the box controller. You will add widgets so users can search for addresses, specify potential restaurant locations, and perform analyses.

17.   Click the + button to add a widget. In the Choose Widget window, choose the Search widget and click OK.

18.   In the Configure Search window, click OK to accept the defaults and close the window.

19.   Click the + button to add another widget. In the Choose Widget window, choose Draw, and click OK.

20. In the Configure Draw window, select Add the drawing as an operational layer of the map, and click OK.

Selecting this option will allow you to use the locations you draw on the map as inputs for your analysis. Otherwise, you must add the potential restaurant locations to your web map as a feature layer or other supported format.

Next, you will add the Analysis widget and configure the necessary tools.

21. Click the Plus button to add a widget. In the Choose Widget window, choose Analysis and click OK.

22. In the Configure Analysis widget, change the widget title to **Restaurant Location Analysis**.

23. Select the Create Drive-Time Areas tool.

24. To the right of that tool under Settings, click the Set tool details button, and select the following settings:

- Set The tool display name as **Step 1: Calculate service area**.
- Clear Show option to use the current map extent.

Using this tool will answer the following question: If most customers are willing to drive up to 15 minutes to reach my restaurant, what geographic area would those 15 minutes cover?

Your analysis will consider two possible restaurant locations, even when they are not included in the current map extent. If you have many candidate locations and only want to analyze points that fall in the map extent, you should leave "Show option to use the current map extent" selected.

25.   Select Enrich Layer, click the Set tool details button, and select the following setting:

  • Set the tool display name as **Step 2: Obtain service area demographics.**

26.   Deselect Show option to use the current map extent.

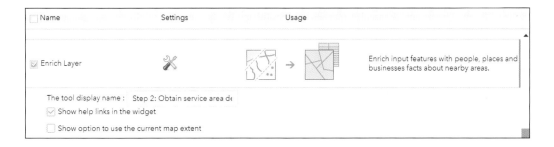

Selecting Enrich Layer allows users to run the ArcGIS Online geoenrichment analysis to find the people, places, and business facts for the input areas.

27.   Select Summarize within, click the Set tool details button, and select the following settings:

  • Set The tool display name as **Step 3: Find block groups for direct mail marketing.**
  • Leave Show option to use the current map extent selected.

This tool will allow users to determine which block groups overlap the service area of a restaurant.

28.   Click OK to close the Configure Analysis window.

29.   Click Save to save your web app.

You have created a web app offering analysis capabilities. Company executives can use this app to select their desired restaurant location.

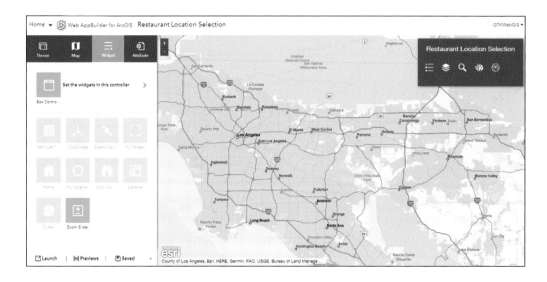

8.2 Perform analysis using the Web AppBuilder Analysis widget

In this section, you will specify two candidate restaurant locations and run the three analysis steps you just configured.

1.   Launch the app you just created.

If you are continuing from the last section, you can click Launch. Otherwise, you can go to your content list, find the app, and view it there.

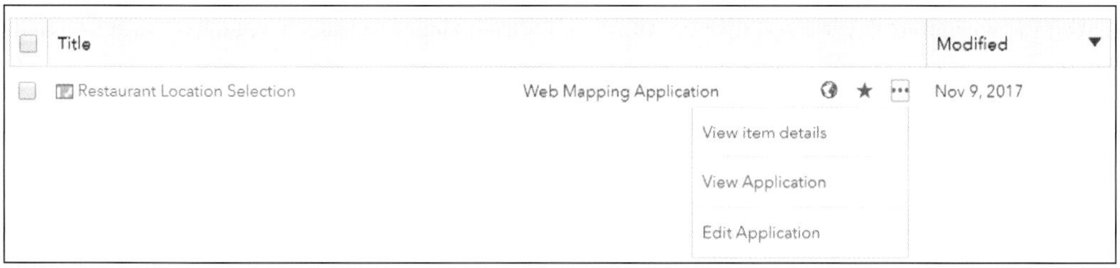

	Title					Modified	▼
☐	🖼 Restaurant Location Selection	Web Mapping Application	🌐	★	⋯	Nov 9, 2017	

View item details

View Application

Edit Application

Next, you will find the first candidate restaurant location and draw a point there.

2. Click the Search button, and in the box at the bottom of the screen, search for 826 W Valley Blvd, Alhambra, CA, 91803.

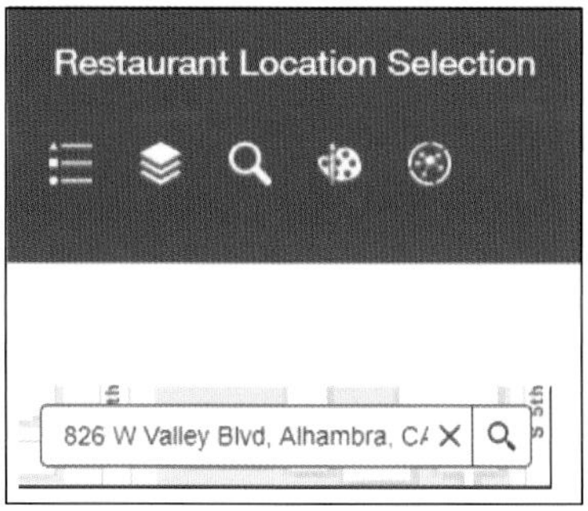

The map will zoom to the location.

3. Click the Draw widget to draw a point at the location of the address.

-   If you cannot find the widget buttons, they probably have minimized to a More button ▤ , and you will click the button to access the tool buttons.

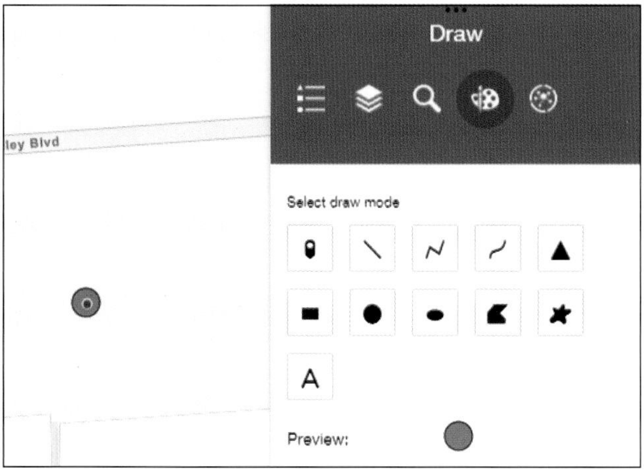

4.   Repeat the previous two steps to add the next candidate location at: **9773 Baseline Rd, Rancho Cucamonga, CA 9173**.

Now that two input locations have been added, you are ready to run the analysis.

5.   Zoom out the map to include both candidate locations.

6.   Click the Restaurant Location Analysis widget.

The Restaurant Location Analysis widget appears, showing the three steps that you configured.

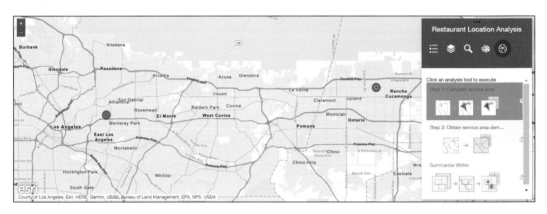

7.   Click step 1, and specify the following parameters:

   - For Choose point layer to calculate drive-time areas around, choose Points from the list.
   - For Measure, choose and set the Driving Time to 15 Minutes.
   - Select Use traffic, choose Traffic based on typical conditions for Friday 6:00 PM.
   - For the Result layer name, specify **Service area of 15m**.
   - Click Run Analysis.

After you run the analysis, you will see that the service areas are added for the two locations. Next, you will get the demographics information for those areas.

8. Click Home to go to the home page of the Analysis widget.

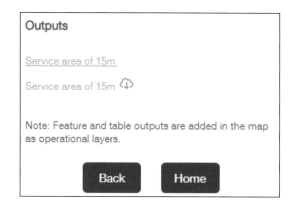

9. Click Step 2. For Choose layer to enrich with new data, choose Service area of 15m from the list.

10. Click Select Variables. In the Data Browser window, perform the following actions:

   - Click Income, and select 2017 Average Household Income (Esri).
   - Click Disposable Income, expand 2017 Disposable Income (Esri), and select 2017 Average Disposable Income (Esri).
   - Click Back and Back.
   - Click Population, select 2017 Total Population (Esri), and 2017-2022 Population: Annual Growth Rate (Esri), and click Back.

   📰 **Note:** The available variables change over time. Feel free to select the variables for more recent years.

   In addition to browsing, you can also search for variables relevant to your needs.

11. Still in the Data Browser window, select the following settings:

   - In the search for a variable name text box, type **Restaurants**, and press Enter.
   - In the resultant variables list, expand 2017 Retail MarketPlace, and select 2017 Retail Sales: Restaurants/Other Eating Places, 2017 Retail Sales Potential: Restaurants/Other Eating Places, and 2017 Leakage/Surplus Factor: Restaurants/Other Eating Places.
   - Click Apply to close the Data Browser window.

12. Change the Result layer name to **Service Area Demographics**.

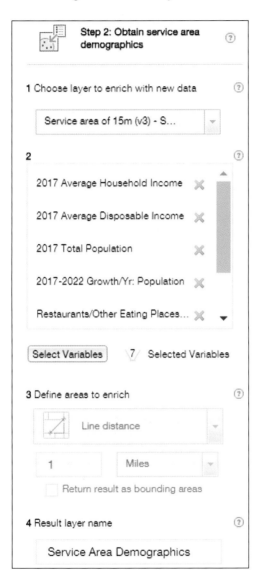

13. Click Run Analysis.

When you complete the analysis, the Output layer appears on the map as an operational layer.

14. Click the Layer List widget, click the More Options button next to the Service Area Demographics - Service Area Demographics layer, and choose View in Attribute Table.

The Attribute table widget opens. Optionally, you can hide some fields by clicking the Options button in the upper-left of the table. Your table values may differ from the figure because the candidate locations you drew on the map might also slightly differ. You might have different variables if you selected different variables in the data browser window.

15. Compare the demographics of the two service areas, and choose your candidate.

For example, you may choose the Rancho Cucamonga candidate (the second point you placed), because its service area has higher income, higher population growth rate, and the area has less restaurant competition.

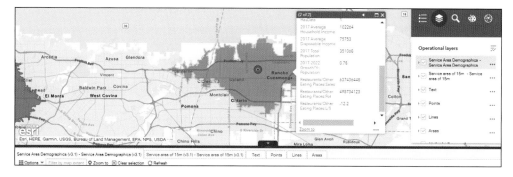

Restaurant competition is indicated by the Restaurants/Other Eating Places: L/S value, which is the leakage/surplus indicator. The indicator value ranges from –100 to 100, with a positive number indicating a leakage, a negative number indicating a surplus, and a large number indicating less competition.

You could also turn on the tapestry segmentation layer in the web map, examine the population profiles of the two candidate areas, and add them into your consideration.

The tutorial is simplified for illustration purposes. In addition to the demographics obtained here, you can consider other factors such as restaurant niche, numbers of vehicles passing by, targeted ages and ethnicity groups, cost to establish the restaurant, and available funding.

Next, you will determine the list of block groups that the restaurant service area overlaps, so that you can distribute flyers for promotional or survey purposes.

To limit the next analysis to the service area of the selected candidate, you can either zoom the map to include only the desired service area, or you can use a filter. This tutorial will use a filter.

16. In the Attribute Table widget, click the Options list, and choose Filter.

17. Click Add a filter expression, select Facility ID (Number), leave the operator as is, click the gear icon, select Unique, and then select 2 from the drop-down list. Click OK.

18. Hide the Attribute table widget by clicking the arrow in the upper middle of the widget.

19. Click the Analysis widget again. Click Home if you do not see the list of three steps.

20. Click Step 3: Find block groups for direct mail marketing.

21. Specify the following parameters:

   • For Choose area layer to summarize other features within its boundaries, choose Service Area Demographics.
   • For Choose layer to summarize, select Choose Living Atlas Analysis Layer.

- In the Choose Living Atlas Analysis Layer window, search for Census BlockGroup, and click choose USA Census BlockGroup Areas in the search result.
- For Choose field to group by, choose FIPS, which is the ID field of the block group layer.
- For Result layer name, set the name as **Overlapping block groups**.

Step 3: Find block groups for direct mail marketing ⑦

1 Choose an area layer to summarize other features within its boundaries ⑦

> Service Area Demographics (v3... ▾

2 Choose a layer to summarize ⑦

> USA Census BlockGroup Areas ▾

3 Add statistics from the layer to summarize ⑦

☑ Sum Area in | Square Miles ▾

Field ▾ | Statistic ▾

4 Choose field to group by (optional) ⑦

> FIPS ▾

☐ Add minority, majority ⑦
☐ Add percentages ⑦

5 Result layer name ⑦

Overlapping block groups

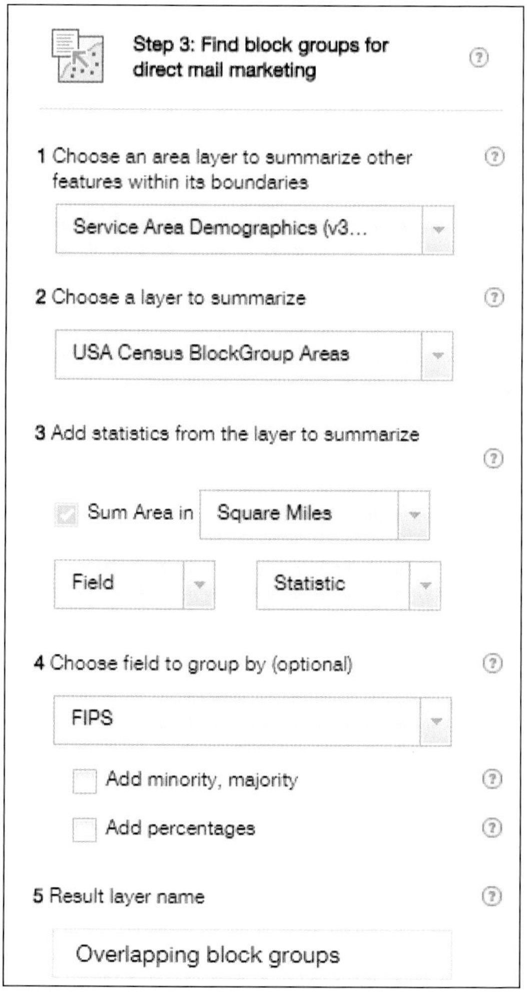

22.   Click Run Analysis.

When you complete the analysis, the result automatically adds to the map.

23.   Click the Layer List widget, click the More Options button next to the Overlapping block groups – GroupBySummary table, and choose View in Attribute Table.

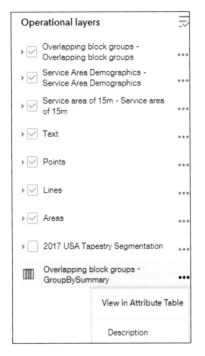

You will see the FIPS of the overlapping block groups in the attribute table.

24.   In the Attribute Table widget, click the Options list, choose Export all to CSV. When this step is confirmed with a pop-up window, click OK and save the CSV file.

Having obtained the list of block group IDs in your selected restaurant service area, you can now arrange direct mail for these block groups. You can also turn on the tapestry layer to understand the socioeconomic status and spending behavior of each block group. This information will help you tailor the survey/marketing messages to each block group.

25.   Close Web AppBuilder.

In this simplified case study, you created a web app that enabled the company to perform restaurant location analysis. You could refine the analysis by obtaining demographic data for more details.

   📃 **Note:** The next four sections teach you how to create and use web tools or geoprocessing services with ArcGIS Enterprise.

8.3 Design a geoprocessing tool (optional)

This book does not focus on designing geoprocessing tools; therefore, it does not discuss many details about ModelBuilder. Instead, the tutorial starts with a mostly completed model.

1. Start ArcGIS Pro.

2. Create a new blank project, set the name as **Factory Site Selection**, and click OK.

3. In the Catalog pane, under the Project tab, look for Folders. Right-click Folders, click Add Folder Connection, navigate to C:\EsriPress\GTKWebGIS, select Chapter8 folder, and click OK.

4. Expand the Chapter8 folder, notice it includes a Lab_Data.gdb, a map document (.mxd), and a toolbox (.tbx).

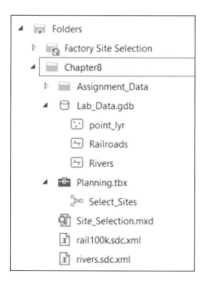

5. In the top-left area of ArcGIS Pro, click the Insert tab, click Import Map, browse to Project > Folders > Chapter 8, select Site_Selection.mxd, and click OK.

The Railroads and Rivers layers display in a map. Next, you will edit the geoprocessing model.

6. In the Catalog pane, find and expand the Planning.tbx file, right-click the Select_Sites model, and click Edit.

7. Review the model elements.

The model includes the following tools:
- **Buffer1** generates a buffer around the location that users click.
- **Clip1** selects rivers that lie within a specified distance from the point of interest.
- **Clip2** selects railroads that lie within a specified distance from the point of interest.
- **Buffer2** generates a buffer around selected railroads.
- **Buffer3** generates a buffer around selected rivers.
- **Intersect3** intersects the two preceding buffers and finds shared areas between them.
- **Dissolve** merges the adjacent little polygons into bigger polygons.

The model is basically complete. You simply must define several input and output parameters.

8. Right-click the InputLocation variable and click Parameter.

The letter P will appear to the upper right of this variable. This letter indicates that the variable is a model parameter. This parameter allows your users to specify new data or values for your model to process.

9.   Right-click the DistanceToLocation variable and click Parameter.

10.   Right-click the DistanceToRailroads variable and click Parameter.

11.   Right-click the DistanceToRivers variable and click Parameter.

12.   Right-click the CandidateSites element and click Parameter.

This will make CandidateSites an output parameter.

13.   Right-click the CandidateSites element and click Add to display.

This will add the result of the model to your map.

14.   Under the ArcGIS Pro ModelBuilder tab, click the Save button to save your model.

15.   Save your project.

Your model is now complete.

8.4 Run the geoprocessing tool (optional)

Your tool must run successfully before it can be shared as a web tool.

1.   Click the Layers tab to make the map window active.

2.   In the Catalog window, find and right-click the Select_Sites model you just completed. Click Open.

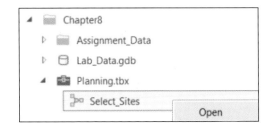

3.   In the Geoprocessing window, specify the following inputs:

- For InputLocation, click the pencil list, and select Points. Click a location on the map near rivers and railroads; otherwise, you might get an empty output.
- Leave the other parameters at their default values.
- Click Run to run the model.

Geoprocessing	▾ ⊹ ✕

| ⊕ | Select_Sites | ≡ |

Parameters │ Environments   ⑦

DistanceToLocation	Linear Unit ▾
90	Kilometers ▾

DistanceToRailroads	Linear Unit ▾
5	Kilometers ▾

DistanceToRivers	Linear Unit ▾
5	Kilometers ▾

CandidateSites
| CandidateSites | 📁 |

ⓘ InputLocation
| | ▾ 📁 ⋰ ▾ |

⋰	Points
⊬	Lines
▣	Polygons
▱	Multipatch

Once the model runs successfully, the model results (CandidateSites) display on the map.

4.   Examine the results in the map.

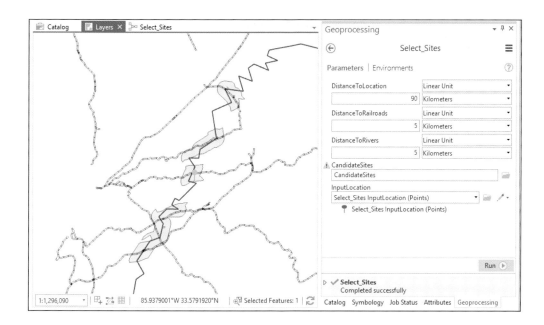

These resultant sites are situated close to the location you clicked on the map and close to rivers and railroads within the distances you specified when you ran the model.

8.5 Publish a web tool and a geoprocessing service (optional)

This section requires ArcGIS Enterprise and a connection to Portal for ArcGIS. Refer to section 6.1 for details on how to add a portal connection.

1. Continuing from the last section, click the Share tab, click the Web Tool, and select Select_Sites.

2. In the Web Tool pane, select the following settings:

    - Set Name as **SiteSelection**.
    - For Data, select Copy all data.
    - For Tags, type **Site Selection** and **GTKWebGIS**.
    - Share with Everyone.

3. In the Web Tool pane, click the Configuration tab, and confirm the Execution Mode is Asynchronous.

4. In the Web Tool pane, click the Content tab, and click the pencil button to the right of Select_Sites.

5. Under the Parameters, select the following settings:

    • Click DistanceToLocation, specify its description as **Enter distance to**

location of interest, and click DistanceToLocation to collapse it.

- Repeat the previous instruction to specify the descriptions for DistanceToRailroads as **Enter buffer distance to railroads**, for DistanceToRivers as **Enter buffer distance to rivers**, for CandidateSites as **Output candidate sites**, and for InputLocation as **Enter location of interest**.

These descriptions will make your tool parameters easier for your users to understand.

6.   Click InputLocation to expand it. For Input mode, select User defined value.

This option allows users to specify input locations by drawing on the map interactively.

7.   In the upper-left corner of the Web Tool pane. click the Back button.

8.   Click the Analyze button, and examine the results in the Messages tab.

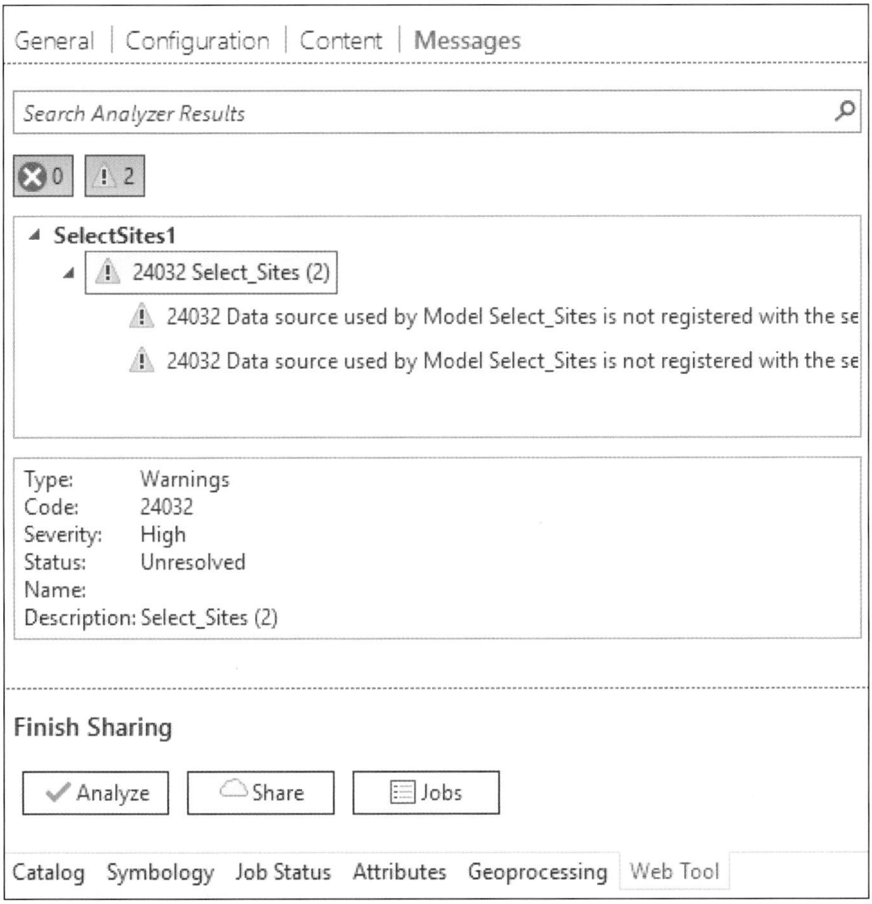

General | Configuration | Content | Messages

Search Analyzer Results

❌ 0   ⚠ 2

▲ SelectSites1
    ▲ ⚠ 24032 Select_Sites (2)
        ⚠ 24032 Data source used by Model Select_Sites is not registered with the se
        ⚠ 24032 Data source used by Model Select_Sites is not registered with the se

Type: Warnings
Code: 24032
Severity: High
Status: Unresolved
Name:
Description: Select_Sites (2)

Finish Sharing

✔ Analyze   ☁ Share   ▤ Jobs

Catalog Symbology Job Status Attributes Geoprocessing Web Tool

The analysis result should find no error. You may see some warnings. As long as there are no errors, you can publish your tool.

9.   Click the Share button.

The publication process may take a couple minutes before you can see a message informing you the tool has been successfully shared.

10.   Save your project.

11.   Click the Manage the web tool link.

You will be directed to the item details page in your Portal for ArcGIS.

12.   On the item details page, on the lower-right corner of the page, find the URL of the web tool, click the copy button, and paste the URL into Notepad or another text editor.

You will use this URL in the next section.

The URL is the REST URL or REST endpoint of your web tool. Web clients access your web tool through this URL.

Next, you will test your web tool.

13.   Click the Open in Map Viewer button.

The tool will be open in the map viewer.

14.   With the tool open, click the Point button for InputLocation, click a location in Alabama, for example, close to the city of Montgomery, and click Run Analysis.

If the point you clicked is not too far from railroads and rivers, the resultant candidate sites will be displayed on the map.

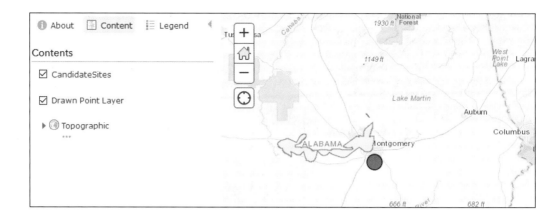

In this section, you published a web tool in Portal for ArcGIS and geoprocessing services in ArcGIS Server behind the scenes. You tested the web tool and made sure it works.

8.6 Use your web tool in Web AppBuilder (optional)

This section does not review the details of Web AppBuilder. Refer to chapter 3 for details.

1.   Continuing from the last section, click Home, and click Content.

If you closed your browser from the last section, you can open a web browser, go to your Portal for ArcGIS, sign in and click Content.

2.   Under the My Content tab, click Create > App > Using the Web AppBuilder.

3.   Specify a Title (for example, **New Factory Site Selection**), Tags (for example, **Site Selection**), optionally a Summary, and then click OK.

4.   Choose the Jewelry Box Theme.

5.   Click the Map tab, and then click Choose Web Map.

You will search for and use an existing web map that shows Alabama's main railways and rivers.

6. Click the Public tab, select ArcGIS Online, type **Alabama Rivers Railroads owner:GTKWebGIS** in the search box, and click the search button. Select the only web map you see, and then click OK.

Choose web map   ✕

My Content	My Organization	My Groups	Public

Alabama Rivers Railroads owner:GTKWebGIS   🔍   Most viewed   ▾   ↻

○ peteamtest2.esri.com   ● ArcGIS Online

Main Rivers and Railroads in Alabama

The map zooms to Alabama and shows the main rivers and railways that your web tool will use.

7. Click the Widget tab, and then click the first placeholder.

Theme	Map	Widget	Attribute

Set the widgets in this controller   >

Header Co...

①

Widget

8. In the Choose Widget window, select the Geoprocessing widget, and click OK.

The Set GP task window appears. You have two ways to set the task. One way is to search in your Portal for ArcGIS, and another way is to specify the service URL directly.

9. Click Portal for ArcGIS, find the SelectSites web tool, and click Next.

You could use the Service URL by clicking Service URL, pasting the URL you copied at step 12 in section 8.5, and clicking Validate.

10. Click the Select_Sites task, and click OK.

The Geoprocessing widget automatically lists the input and output parameters of your task.

11. In the Input parameter list, click InputLocation. Under Input feature by, select Interactively drawing on the map.

Optionally, you can click each of the input parameters and change the labels and tooltips for better readability.

12. Click OK to close the Configure Geoprocessing window.

13. Click Save to save your app.

Your widget is now configured. Next, you will test the widget in your web app.

14. Click Launch to open your app.

The geoprocessing widget opens by default.

15. For the input parameters, use the default values or specify new values, for example, **50** kilometers for DistanceToLocation, **8** kilometers for DistanceToRailroads, and **9** kilometers for DistanceToRivers.

16. Under **InputLocation**, click the Point button. Click a location close to rivers and railroads on the map to specify your area of interest, and then click **Execute**.

After a few moments, a set of candidate sites for the new factory displays on the map.

8.7 Perform big data analysis using ArcGIS Enterprise (optional)

Because of the size of data to be downloaded and the system required to support the analysis, instructors may not have time to teach this section in class, and readers may not have the environment to run the analysis. If this situation applies to you, the next section is for reading only, so that you can get a basic understanding of the workflow for big data analysis.

1. Open a web browser, go to **http://arcg.is/2yR26N5** or search for Tutorial: Run a GeoAnalytics tool.

2. Read through the page and review the following figures while reading the corresponding sections.

    - Refer to the next figure for creating a big data file share.

- Refer to the next figure for editing a big data file share.

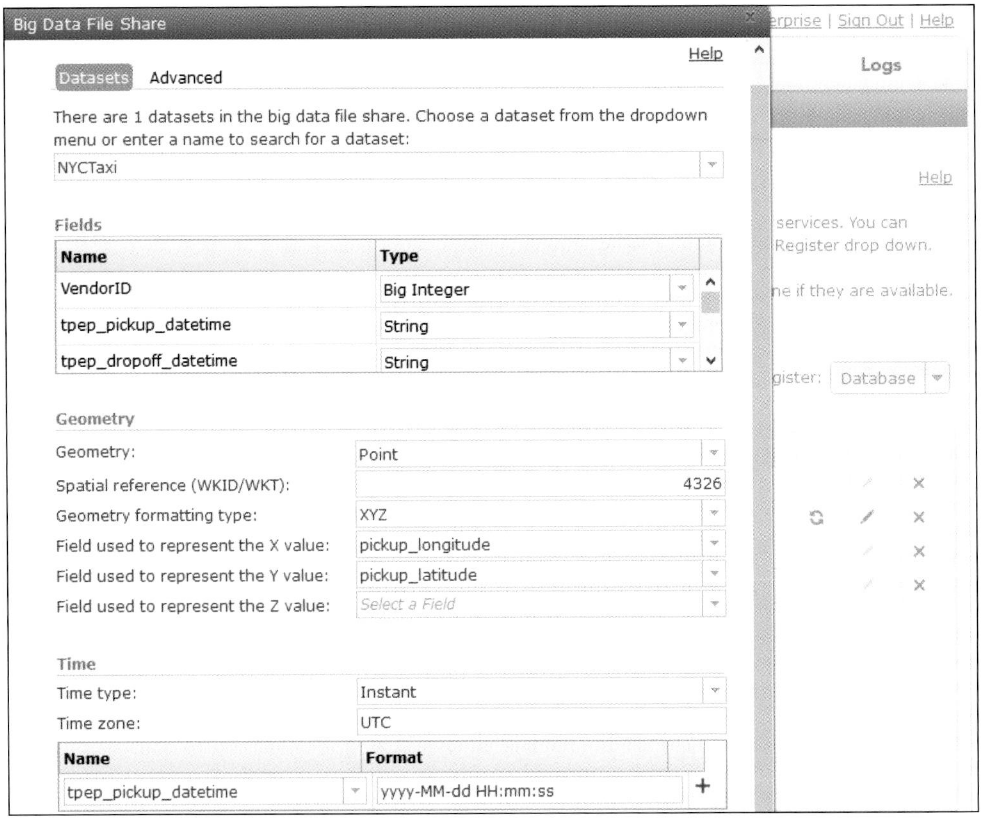

- Refer to the next figure for selecting the Aggregate Points tool.

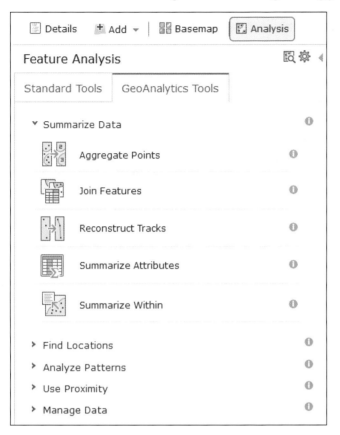

- Refer to the next figure for selecting the big data file share.

- Refer to the next figure for specifying parameters for the Aggregate Points tool.

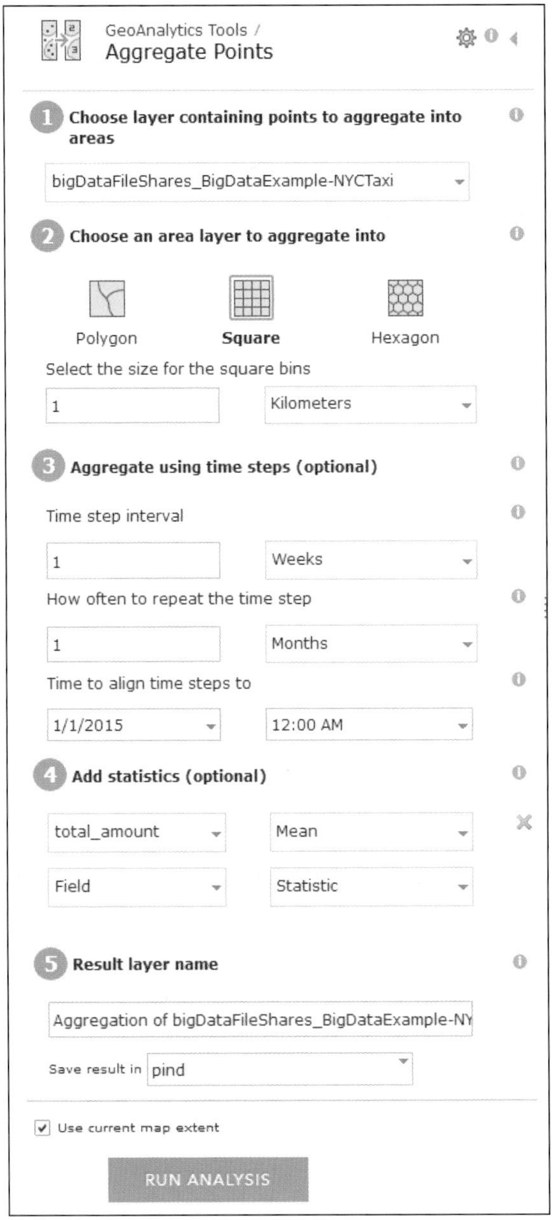

- For the analysis result, refer to the New York City taxi figure earlier in this chapter.

In this tutorial, you employed ArcGIS standard analysis tools using Web AppBuilder in sections 8.1–8.2, created and used your own web tool using ArcGIS Enterprise in sections 8.3–8.6, and reviewed the steps to perform big data analysis using GeoAnalytics Server in section 8.7.

QUESTIONS AND ANSWERS

1. I am a publisher in my ArcGIS Enterprise. I can neither publish geoprocessing services from ArcMap nor share web tools in ArcGIS Pro. Why?

   Answer: With ArcGIS Enterprise 10.4 and later, only administrators can publish geoprocessing services and web tools by default. This limitation provides greater security for your ArcGIS Enterprise. Your ArcGIS Enterprise administrator can create a custom role with the Publish web tools privilege and assign you this custom role. Your ArcGIS Enterprise administrator can also choose to allow all publishers to publish geoprocessing services. To do so, follow these steps:
   - Open the ArcGIS Server Administrator Directory and log in with an account that has administrative privileges to the site. The URL to the Administrator Directory is typically similar to **https://gisserver.domain.com:6443/arcgis/ admin**.
   - Click System > Properties > Update.
   - Enter **{"allowGPAndExtensionPublishingToPublishers": true}**.
   - Click Update.

   ArcGIS Server Administrator Directory

   Home > system > properties > update

   Update Server Properties

   System Properties

   Properties (in JSON format):* {"allowGPAndExtensionPublishingToPublishers": true}

   Format: HTML ▾

   Update

2. ArcGIS Online analysis costs credits. Can I determine how many credits my analysis will cost before I run the analysis?

Answer: You will find a Show credits link next to the Run Analysis button in the Perform Analysis pane of the portal map viewer, as well as in the Analysis widget of Web AppBuilder. Click the Show credits link to see the credits that will be used.

3. Can I combine or chain multiple ArcGIS Online analysis tools and run the tools automatically without having to run each tool manually?

Answer: Yes. You can write your own programs using JavaScript, ArcPy, ArcGIS API for Python, and other languages. You can use the output of one tool as the input for another tool, and thus can chain multiple tools.

4. I deleted my web tool from Portal for ArcGIS, but I still can't publish a web tool with the same name as the web tool I just deleted. Why?

Answer: A web tool in Portal for ArcGIS references a geoprocessing service in ArcGIS for Server. You need to delete the geoprocessing service in ArcGIS Server. You can do so using ArcGIS Server Manager.

5. As I consider ModelBuilder or Python, which one should I use to author my model or tool?

Answer: The answer varies depending on what you want to accomplish and on your current skillsets.
    Use ModelBuilder if you are new to both ModelBuilder and Python and must finish the work quickly.
• ModelBuilder is much easier and quicker to learn than Python.

- ModelBuilder excels at visually and intuitively depicting the workflow for most tasks that you want to accomplish.
- Even if you prefer to learn Python, ModelBuilder can be a better place to start. ModelBuilder can export some models to Python scripts, so the skeleton of the script is already written for you.

If you already know or have more time to use Python, go ahead and use Python.

- Python is a scripting language. You can accomplish more complex workflows and exert more fine-grained control. For example, simple text manipulation is difficult using ModelBuilder, whereas Python makes the task easier.
- With additional libraries or modules, Python can better integrate with other software tools, such as Microsoft Excel, statistical Package R, deep learning and artificial intelligence APIs, or procedures in a relational database management system (RDBMS).
- In addition to using your tool to create a geoprocessing service, you may also want to run your Python script outside ArcGIS Pro or schedule it to run at a certain time.

6. I can publish a Python/ArcPy script as a web tool. Can I publish a script written in ArcGIS API for Python as a web tool too?

    Answer: No, not yet. As for now, ArcGIS API for Python is intended for use as a client to perform GIS visualization, analysis, and administrative tasks.

ASSIGNMENTS

Choose one of the two options.

Assignment 8A: Use ArcGIS Online analysis to select your new home.

You are in the market for a new home. You are considering many factors, such as the distance between your home and work or your children's schools, neighborhood income, median house value, median family income, unemployment rate, and crime rate. List your consideration factors and criteria, and perform analysis in ArcGIS Online map viewer to help you narrow the area in which you would like to buy your new home.

Data:

- Provide the locations of you and your family members' offices or schools by yourself.
- Use ArcGIS Online Living Atlas layers and data browser variables.

Tips:

- You can add your and your family members' work or school locations using a CSV file, or you can draw the locations interactively.
- Select three tools and chain them (in other words, using the result of one tool as the input of another), if appropriate.

You will probably need the Enrich layer tool. Use the data browser to see what variables will help you choose your new home.

- Use Web AppBuilder to create a web app so that others can use it to decide where to buy new homes.

What to submit:

- The URL of the web app you created using Web AppBuilder
- Screenshots of your analysis processes and your final choice of neighborhood areas

Assignment 8B: Create a web app to clip, zip, and ship GIS data.

A survey and mapping bureau would like to improve its data-sharing workflow. Instead of extracting and copying data manually to serve its customers, the bureau wants to automate the workflow with a web app.

Data: The data is located at **C:\EsriPress\GTKWebGIS\Chapter8\Assignment_Data**. It contains the following data:

- **data.gdb**, which has the Earthquakes and Hurricanes layers that your users can download.
- **natural_disasters.mxd**, which displays these data layers.

ExtractData.tbx, which contains ExtractData, a to-be-completed model for clipping, zipping, and shipping data.

Requirements: This web app should allow its users to select the layers they need, draw the area of interest, select the desired data projection and format, and have the data clipped and zipped for them to download.

Tips:

- Display Natural_disasters.mxd in ArcGIS Pro.
- Edit the ExtractData model, and set Layers to Clip, Feature Format, Area of Interest, and Output_zip File as model parameters.
- Run the ExtractData model and share the result of the model as a web tool.
- Create a web app using Web AppBuilder for ArcGIS:
    - Use the web map you created in the chapter 6 tutorial. This web map displays the Earthquakes and Hurricanes layers so users can see the locations of the earthquakes and hurricanes.

Use the Geoprocessing widget and configure it to point to the ExtractData task in your web tool.

What to submit:

- The URL or REST endpoint of your web tool
- The URL to your web app

Resources

"ArcGIS API for Python," https://developers.arcgis.com/python.

"Build a Model to Connect Mountain Lion Habitat, " http://learn.arcgis.com/en/projects/build-a-model -to-connect-mountain-lion-habitat (or http://arcg.is/2A5EIeP).

"Create a model tool," http://pro.arcgis.com/en/pro-app/help/analysis/geoprocessing/basics/create-a -model-tool.htm (or http://arcg.is/2hwyGbM).

"Esri UC 2017: ArcGIS API for Python," Rohit Singh and Mansour Radd, https://www.youtube.com/ watch?v=b_s31fujHT0 (or http://bit.ly/2jyu0qg).

"Geoprocessing widget in Web AppBuilder for ArcGIS," http://doc.arcgis.com/en/web-appbuilder/create -apps/widget-geoprocessing.htm (or http://arcg.is/2ARheXX).

"Get Started with Insights for ArcGIS," https://www.esri.com/training/catalog/ 5899f4b295dd882431ce77b7/get-started-with-insights-for-arcgis/ (or http://arcg.is/2fGcCLi).

"I Can See for Miles and Miles," http://learn.arcgis.com/en/projects/i-can-see-for-miles-and-miles.

"Insights for ArcGIS," http://www.esri.com/products/arcgis-capabilities/insights (or http://arcg.is/ 1Rjwumb).

"Insights for ArcGIS: An Introduction," Art Haddad and Linda Beale, https://www.youtube.com/watch?v= 36ODQxIizKA (or http://bit.ly/2zLFUmz).

"Perform analysis in ArcGIS Online," https://doc.arcgis.com/en/arcgis-online/analyze/perform-analysis .htm (or http://arcg.is/2z85O59).

"Perform analysis in Portal for ArcGIS," http://server.arcgis.com/en/portal/latest/use/perform-analysis .htm (or http://arcg.is/1sAC7Bw).

"Perform big data analysis using ArcGIS GeoAnalytics Server," http://server.arcgis.com/en/server/latest/get-started/windows/perform-big-data-analysis.htm (or http://arcg.is/2pxylJO).

"Python Scripting for Geoprocessing Workflows," https://www.esri.com/training/catalog/5763042c851d31e02a43ed84/python-scripting-for-geoprocessing-workflows (or http://arcg.is/2z9yqL2).

"Quick tour of authoring and sharing web tools," https://pro.arcgis.com/en/pro-app/help/analysis/geoprocessing/share-analysis/quick-tour-of-authoring-and-sharing-web-tools.htm (or http://arcg.is/2j0pb53).

"Use the analysis tools in ArcGIS Online," http://doc.arcgis.com/en/arcgis-online/use-maps/use-analysis-tools.htm (or http://arcg.is/2A5zRdB).

"What Is ArcGIS GeoAnalytics Server?" http://server.arcgis.com/en/server/latest/get-started/windows/what-is-arcgis-geoanalytics-server-.htm (or http://arcg.is/2zMXchl).

"What Is a web tool?" https://pro.arcgis.com/en/pro-app/help/analysis/geoprocessing/share-analysis/what-is-a-web-tool.htm (or http://arcg.is/2jA6GZn).

"Where Does Healthcare Cost the Most?" http://learn.arcgis.com/en/projects/where-does-healthcare-cost-the-most (or http://arcg.is/2mMgkGH).

CHAPTER 9

Image service and online raster analysis

By Dr. Jie Chang

You have used map services published from imagery in previous chapters. For example, the World Imagery basemap is one of the most common map services used for reference to provide users with realistic, visual information. Imagery can also be published as image services and used for both visual and spatial analysis. For example, the multispectral Landsat image service can be viewed with different band combinations. The agriculture band combination (bands 6, 5, 2) highlights differences between crop types and is especially useful for agriculture applications. The multispectral Landsat image service can also be used to generate a Normalized Difference Vegetation Index (NDVI) to provide the information on the health of vegetation. In this chapter, you will learn the basic concepts of raster and imagery and how they are managed in ArcGIS. You will also learn the concepts of image services, how to publish image services, and how to perform raster analysis online.

Learning objectives

- *Understand the concepts of raster, imagery, and image services.*
- *Familiarize yourself with image layers in the Living Atlas.*
- *Prepare and publish image services.*
- *Perform raster analysis online using raster function templates.*
- *Create web apps using image services.*

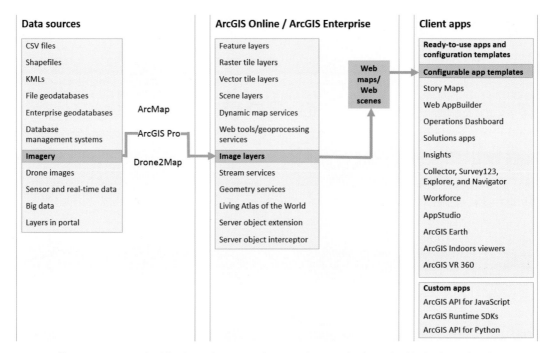

ArcGIS offers many ways to build web applications. The green lines in the figure highlight the technology presented in this chapter.

Raster and imagery

Raster is a data format that consists of a matrix of cells (or pixels) organized into rows and columns (or a grid) where each cell contains a value representing information. The cell value can represent continuous, numeric data like elevation values and categorical data like land cover categories. The cell size determines the spatial resolution of the raster data. A cell size of 10 meters means that each cell covers an area of 10 × 10 meters. The smaller the cell size, the higher the resolution of a raster and, thus, the greater the detail. Raster format is organized as one or more bands. A single-band raster only contains a single matrix of cell values. An example of a single-band raster is a DEM. Each cell in a DEM contains only one value representing surface elevation. A multiple band raster contains multiple, spatially coincident matrices of cell values. Multispectral and hyperspectral imagery have multiple bands, with each band typically containing values within a specific wavelength interval of the electromagnetic spectrum.

361.4	359.1	355.4	349.1	345.5	338.2
353.2	353.4	351.9	348.2	344.7	340.5
348.7	346.6	346.4	344.8	343.9	344.2
350.1	347.9	345.3	345.4	346.1	346.2
353.9	353.2	352.8	351.2	349.4	346.3
358.1	359.8	359.2	357.8	354.4	357.3

1	1	1	1	1	3
1	1	1	1	1	3
1	1	1	2	2	3
1	1	1	2	2	3
2	2	2	2	3	3
2	2	2	2	3	3

Raster cell values represent elevation values (left) and land cover categories (right).

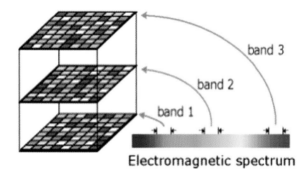

Each band of a multispectral image contains values within a specific wavelength interval of the electromagnetic spectrum.

Imagery in ArcGIS is used to describe all forms of raster data including satellite imagery, aerial photographs, drone-acquired images, DEMs, scanned maps, video data, and thematic raster data, such as land classification or the grid-based output from an analysis or interpolation process. Basically, imagery in ArcGIS refers to any raster or cell-based data.

Imagery is often used as a basemap for reference. Orthorectified satellite imagery and aerial photographs are used as basemaps to provide visual information. Imagery is also used for various spatial analyses. For example, multispectral imagery is classified into different land use/cover categories; DEMs are used in various surface analyses like terrain analysis, visibility analysis, hydrologic analysis, and so on. Multiple thematic raster data is used in weighted overlay analyses to solve multicriteria problems such as site selection and suitability models.

Recent advancements of remote sensing technologies provide collections of imagery with increasingly higher resolution and accuracy. WorldView-4 satellite imagery, for example, whose spatial resolution reaches 0.31 meters, is comparable to that of aerial photographs. Lidar-derived DEMs can accurately model the minor relief of the ground surface. Drone-acquired images have become an important source for orthoimages, DEMs, and 3D models.

Various advanced technologies are also used in processing and analyzing imagery to extract information. AI shows promise in high-resolution image classification. Recently, Microsoft has cooperated with Esri to apply AI in high-resolution image classification. They use the Azure GPU-powered virtual machine, a Microsoft technology called Cognitive Toolkit, and ArcGIS software to train the deep learning algorithm using the high-resolution land classification map of the Chesapeake watershed. The goal is to produce a single land classification model that can be used to classify land cover not only in the Chesapeake Bay area but also in other US locations.

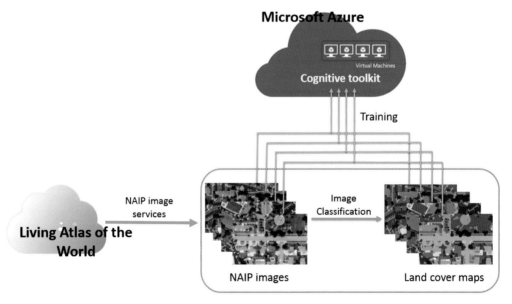

The NAIP image service is used to provided quick access to the NAIP imagery. The selected images are first classified as land cover maps (refined manually when necessary). The images and corresponding land cover maps are then used to train the Cognitive Toolkit running on Azure.

Raster dataset and mosaic dataset

- **Raster dataset:** A raster dataset is the basic raster data storage model in ArcGIS. A raster dataset is any valid raster format organized into one or more bands. ArcGIS supports more than 70 different file formats for a raster dataset, including TIFF, JPEG 2000, Esri GRID, and MrSid. Raster datasets can be stored as files on disk or within a geodatabase.
- **Mosaic dataset:** A mosaic dataset is a collection of raster datasets stored as a catalog and viewed or accessed as a single mosaicked image. A mosaic dataset is stored within a geodatabase, but the raster datasets it references can remain in their native format on disk or exist in the geodatabase. A mosaic dataset is designed for a large collection of raster datasets.
  - Mosaic datasets reference raster datasets without copying them into the geodatabase, thus avoiding data duplication and saving space on the disk. In addition, the source pixels are never altered with any operations in the mosaic datasets.
  - Mosaic datasets can manage raster datasets in different projections, resolutions, pixel depths, and number of bands.
  - Mosaic datasets are dynamically mosaicked. They use the mosaic method to sort the raster and use the mosaic operator to solve the overlaps. During zooming and panning, the raster datasets that meet the given extent and visible range request will be selected and sorted based on the user-specified mosaic method to create a virtual seamless mosaic on the fly.
  - Mosaic datasets use raster functions to conduct quick on-the-fly processing.
  - Mosaic datasets use overview to display quickly and seamlessly at all scales.
  - The mosaic dataset contains metadata of the source images, such as location, sensor type, image acquisition data, and so on. Images can be easily searched and retrieved from the mosaic dataset using attribute and location queries.
  - Mosaic datasets provide an automatic updating mechanism. New images in the data folder can be automatically added to the mosaic dataset using the same raster type used in the mosaic dataset through a synchronize operation. Existing images in the mosaic dataset can also be updated if the sources or image properties have changed.

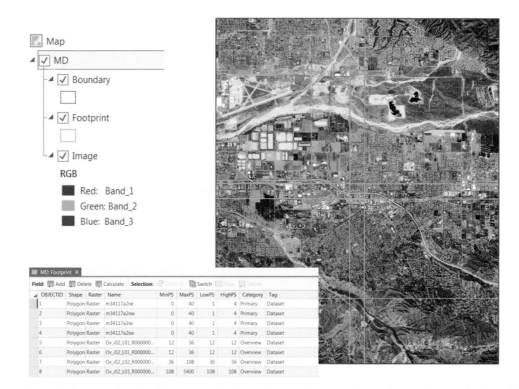

A mosaic dataset is displayed as a composite layer of Footprint, Boundary, and Image layers, which provides a virtual mosaic view of rasters within the mosaic dataset. The Footprint table contains a shape field, a raster field, a MinPS field, a MaxPS field, and other metadata fields. The shape field stores the image footprints. The raster field stores the references to the image data and the raster functions. The MinPS and MaxPS fields define the visible ranges of the rasters.

Image service

Imagery can be shared as an image service using ArcGIS Enterprise. An image service provides access to raster data through a web service. An image service is useful because you can use it as if you're accessing the source raster data directly. You can use it as an image for visual analysis or as a basemap. You can access the pixel values so an image service can be used as the source in any image analysis, or just to identify the pixels. And if you're serving the mosaic dataset, you have access to the attribute tables so you can perform queries and even download the source images.

Imagery can also be published as other web services, such as a map service. Publishing imagery to an image service or a map service depends on the purpose of your application. In general, if the purpose of your imagery is only for visualization as a basemap and you expect moderate-to-high amounts of traffic on your site, you should attempt to publish your imagery as a map service. If you want to expose analysis and manipulation of the imagery, and you don't think the concurrent requests will overwhelm your server, you should publish your imagery as an image service.

Image services available in the Living Atlas

The collections of image services in the Living Atlas are most often found in the Imagery and Landscape categories. Below are brief descriptions of some image services:

- **Multispectral Landsat:** This worldwide dynamic image service provides access to Landsat 8 Operational Land Imager (OLI) multispectral 8 band scenes, acquired through the Landsat program jointly managed by the National Aeronautics and Space Administration (NASA) and the USGS. This service includes various server-side functions that render different band combinations and indices.

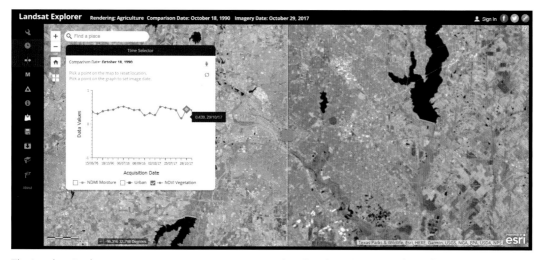

The Landsat Explorer app compares two images acquired on October 18, 1990, and October 29, 2017, near Dallas, Texas. The images are rendered in the agriculture band combination, which highlights the agriculture in bright green. The profile shows the NDVI trend for the selected point (red dot). This app accesses the Landsat image service running on the Amazon Web Services Cloud. The app provides quick access to the different band combinations and indices. The app allows image comparison and change detection. The app also provides a spectral profile or temporal profile for the selected point.

- **Terrain:** This image service provides online access to global collections of multiresolution, multisource elevation data. This collection includes data with resolutions ranging from about 1,000 meters to 0.5 meters. This image service provides numerous server-side functions for visualization and analysis.

- **NAIP Imagery:** The imagery is captured through the National Agriculture Imagery Program (NAIP) and made available by USDA Farm Service Agency. NAIP acquires aerial imagery during the agricultural growing seasons in the continental US. The image layers include 4-band NAIP imagery (RGB and near infrared) with 1m and better resolution since 2010. This image service provides server-side functions including natural color, false color, and NDVI.

- **World Land Cover 30m BaseVue 2013:** MDA's BaseVue 2013 is the most up-to-date global, land use/land cover (LULC) product at 30-meter resolution. It is derived from roughly 9,200 Landsat 8 images. LULC is classified into 16 classes.

- **GLDAS layers:** Global Land Data Assimilation System (GLDAS) layers include five image layers: monthly evapotranspiration layer, monthly precipitation layer, monthly runoff layer, monthly snow pack layer, and monthly soil moisture layer. These layers are time-enabled image services showing outputs from the Noah land surface model, which runs at 0.25-degree spatial resolution using satellite and ground-based observational data from NASA's GLDAS-2.1. The model is run with 3-hour time steps and aggregated into monthly averages from January 2000 to the present. These image layers provide server-side functions for visualization and analysis.

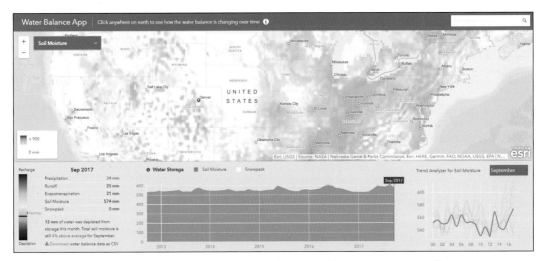

The Water Balance app ingests GLDAS layers. The app shows monthly precipitation, runoff, evapotranspiration, soil moisture, and snowpack near Denver, Colorado, in September 2017. The app shows how much recharge or depletion occurred during this month and how that amount compares to the average. The app also shows the trend of the chosen variable in the same month during other years.

- **Sea Surface Temperature:** This time-enabled layer shows the average sea-surface temperature during the map's time extent. The sea-surface temperature dataset is calculated from satellite-based microwave and infrared imagery by the Naval Oceanographic Office. This data is optimally interpolated to provide a daily, global map of the midday (12:00 p.m.) sea-surface temperature.

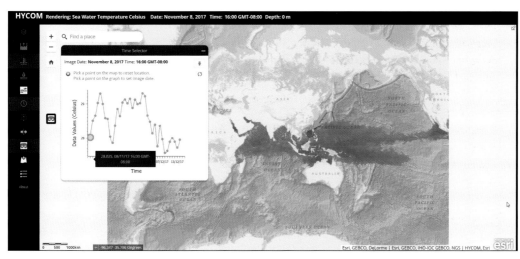

The HYCOM web app uses the sea-surface temperature image layer. The orange color shows the areas where sea-surface temperatures (depth is 0 meter) are between 27.7 and 32 degrees Celsius on November 8, 2017, at 16:00 GMT-08:00. The profile shows the temporal changes of temperature at a specific location (indicated by the green arrow).

A list of image layers for the Living Atlas of the World

Image service	Data source	Coverage	Resolution	Server-side functions
Multispectral Landsat	Landsat Program	Global	30 m	Different band combinations (natural color, color infrared…) and indices (NDVI, NDMI, NBR)
Terrain	Multisource	Global	0.5 m–1,000 m	Slope, Aspect, Hillshade…
NAIP Imagery	The National Agriculture Imagery Program (NAIP)	Continental United States	1 m and 0.6 m	Different band combinations (natural color and false color) and NDVI
World land cover 30 m BaseVue 2013	MDA	Global (extruding Antarctica)	30 m	
GLDAS layers (evapotranspiration, precipitation, runoff, snow pack, and soil moisture)	Global Land Data Assimilation System (GLDAS)	Global	0.25 degree	Calculate anomaly, cartographic render, analytic render…
Sea water temperature	Naval Oceanographic Office	Global	11 km	Cartographic, convert to Fahrenheit

Prepare image service

You must consider many factors before publishing image services, such as the kind of data that can be published as image services and where to store the data to be published. You must understand the various functions and capabilities of image services and consider creating caches to improve the performance. You should also consider applying raster functions to image services to perform online raster analysis.

- **Data source:** The source for the image service can be a raster dataset, a mosaic dataset, or a layer file referencing a raster dataset or a mosaic dataset. Please note that your ArcGIS Server must be licensed with an image server role to publish data related to mosaic datasets or mosaic functions.
- **Data store:** The data can either be copied to or registered with the server. Registering data to the server can avoid data duplication and save space on the server.
- **Parameters:** The parameters control how the raster data is made available as an image service and how clients can interact with the image service. Some parameters apply to all image services regardless of whether their data source is raster datasets or mosaic datasets, such as the default resampling method and allowed compression methods. Other parameters only apply when the input for an image service is a mosaic dataset, such as whether downloading is allowed.
- **Capabilities:** Image service capabilities determine how clients can access the image service. An image service is always published with imaging capabilities, allowing clients to connect to it using an ArcGIS Server connection or via REST. The image services can be displayed and analyzed. Optionally, an image service can be published with the OGC WMS or WCS capabilities. Other applications that meet the requirements to access WMS or WCS can connect to the image service. WCS allows clients to access image data, so the image service can be used for display or analysis. WMS allows clients to access an image as a picture, so the image service can only be used for display.
- **Caches:** Image services can be cached to improve the performance of image services in client applications. When you cache an image service, the server pre-generates a collection of image tiles at multiple predefined scale levels or pixel sizes, similar to map service caching. The cached image service can display an image rapidly each time a request is made because ArcGIS Server does not have to generate the image dynamically.

Workflow to publish image services

The following workflow is typical for publishing image services using ArcGIS Pro:

1. Prepare your data. Create pyramids and calculate statistics for raster datasets, and build overviews and calculate statistics for mosaic datasets. Create raster function template files if you want to apply on-the-fly processing for the image services.

2. Set your Portal for ArcGIS as Active Port.
3. Because image services can only be published from the Catalog pane, create a folder connection in the Catalog pane to connect to the folder containing the data you want to publish.
4. Choose referencing data with the server or copying data on the server. Referencing data with the server is recommended to publish a large raster dataset or mosaic dataset to avoid data duplication and save space on the server.
5. Fill out service items and configure web layer properties.
6. Analyze. You must solve all the problems before publishing the service.
7. Publish the image service.

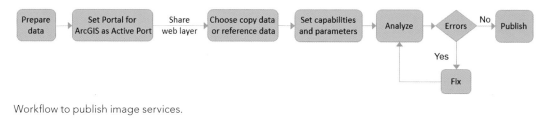

Workflow to publish image services.

Raster function and online analysis

Raster function

A raster function is a pixel-based mathematical model that defines operations directly with the pixels of raster data. Only pixels at the required resolution and extent are processed, which reduces the processing time by avoiding processing pixels that are not needed. For example, only pixels that are visible on your screen are processed with local functions. Running raster functions does not create processed files on disk, so the processes can be applied quickly. The raster function provides a foundation for dynamic image processing. The following is a list of commonly used raster functions:

A list of commonly used raster functions

Function	Description
NDVI	Calculates the Normalized Vegetation Differential Index (NDVI) values using the red and near-infrared bands
Pansharpen	Enhances the spatial resolution of a multiband image by fusing it with a higher-resolution panchromatic image
Hillshade	Creates a grayscale 3D representation of the terrain, with the sun's relative position taken into account for shading the surface
Shaded relief	Is similar to Hillshade but uses color for better cartographic representation
Colormap	Transforms the pixel values to display the raster data as a grayscale or an RGB image, based on a color map
Contrast and brightness	Adjusts the difference between colors and overall lightness of the image
Stretch	Enhances an image by changing properties, such as brightness, contrast, and gamma through multiple stretch types
Remap	Groups pixel values together and assigns the group a new value
Clip	Extracts or excludes an area in a raster according to the extent or the polygon boundary
Slope	Calculates the rate of change from one pixel value to its neighbors
Aspect	Identifies the downslope direction of the maximum rate of change in value from each cell to its neighbors
Arithmetic	Uses the pixel values to calculate mathematical operations on overlapping rasters
Weighted overlay	Overlays several rasters using a common measurement scale and weights each according to its importance

Raster functions can be chained together to apply more complex processing to raster, which is similar to the process chains used in the ArcGIS ModelBuilder. The advantage of using a composite function is that it does not need to output intermediate results; instead, it can produce the final result directly and quickly.

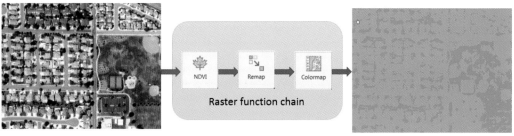

Multispectral Imagery   Raster function chain   Vegetation Map

This composite function chains NDVI function, Remap function, and Colormap function. The multispectral image is classified into vegetation and non-vegetation based on NDVI values, and the classification results are displayed in two colors.

Server-side on-the-fly processing

Image services support server-side processing with raster functions. When you publish an image service, you can define processing (using raster functions) that is applied by the server on the fly. Server-side processing using raster functions has the following advantages:

- It is on-the-fly processing. The processing is fast because only the requested pixels are processed and the results are not written on the disk.
- It is server-side processing. Web applications are often unable to perform processing on image services and they rely on server-side processing.
- An image service can be processed using different methods when it is published with different raster function template (.rft.xml) files.

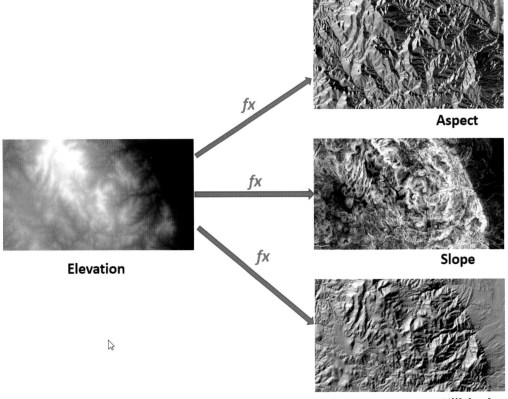

An elevation raster is published as a single image service but can be processed with different methods.

```
In [16]:    from arcgis.raster.functions import *

            land_water = stretch(extract_band(landsat_lyr, [4, 5, 3]),
                                 stretch_type='PercentClip',
                                 min_percent=2,
                                 max_percent=2,
                                 dra=True,
                                 gamma=[1, 1, 1])
```

```
In [17]:    map2 = gis.map("Pallikaranai", zoomlevel=13)
            map2
```

The image service can also be processed on the fly using Python API. The code uses the Extract Band function to create a band combination of 4, 5, 3, and applies the Stretch function to get the land-water boundary visualization that makes it easy to see where land is and where water is.

This tutorial

This tutorial includes two parts. The second part requires ArcGIS Enterprise and is optional.

Part 1 (sections 9.1–9.2): Create a web app in ArcGIS Online using Living Atlas image services to show the change of vegetation in Redlands, California, during 2012 and 2016.

Data: USA NAIP Imagery: Natural Color image service in Living Atlas of World.

Functional requirements: Your web app should display the following capabilities:

- Displays NAIP imagery with true color, false color, and NDVI color.
- Compares two overlapping NAIP images acquired at different dates.
- Calculates NDVI for two overlapping NAIP images and calculates their difference to show the changes in vegetation.

System requirements:
- A publisher or administrator account in an ArcGIS Online organization.

Part 2 (sections 9.3–9.4): Publish an image service to ArcGIS Enterprise and create a web map using the image service.

Data: 4-band NAIP image of Redlands, CA (2012).

Functional requirements: Your web map should display the following capabilities:
- Can display the image with different band combinations (for example, true color, false color).
- Can display vegetation distribution using the NDVI Colorized function.

System requirements:
- ArcGIS Pro 2.1 and higher.
- A publisher or administrator account in ArcGIS Enterprise 10.6 or higher.

9.1 Create a web map in ArcGIS Online using a Living Atlas image service

In this section, you will create a web map in ArcGIS Online using an image service from the Living Atlas. The image service you will use in this section is called USA NAIP Imagery: Natural Color. This image service is a collection of NAIP imagery acquired during different years in the US. This source of this image service is not a raster dataset, but a mosaic dataset.

1. Sign in with your ArcGIS Online account. Click Content.

2. In My Content, click Create. From the drop-down menu, select Map.

3. In the New Map page, perform the following actions:
   - For Title, specify **NAIP Imagery**.
   - For Tags, specify **NAIP** and **GTKWebGIS**.
   - For Summary, input **A web map with Living Atlas image service USA NAIP Imagery: Natural Color**.
   - For Save in folder: select Chapter9.
   - Click OK.

A map opens with the default Topographic layer.

4. Click Add. From the drop-down menu, select Browse Living Atlas Layers.

5. In the Living Atlas pane, search NAIP.

6. In the search results, click the Add button under USA NAIP Imagery: Natural Color to add the layer to the map. Zoom in the layer to its full extent. Click the Back button to return to the Contents pane.

The USA NAIP Imagery: Natural Color layer is added to the map and displayed in natural color.

7. In the Contents pane, for USA NAIP Imagery: Natural Color, click More Options, and click Rename.

8. Change the layer name to NAIP Imagery. Click OK.

The layer name is changed to NAIP Imagery.

9. Click the image. A pop-up window appears, which reads: "You've selected the Overview raster." Close the pop-up window.

The image layer is currently at its full extent. The image you see is the overview, not the source image.

10. In the Search box, search **Esri**. In the search results, click Esri - 380 New York St, Redlands, CA, 92373, USA.

The map zooms to Esri.

11. Close the Search result window. Click the image.

A pop-up window appears, which reads, "You've selected the Primary raster."

12. Close the pop-up window.

The image you see now is the source image. The pop-up window also shows the source image's information. The cyan line is the source image's footprint. The image displaying was acquired on May 2, 2016.

13. In the Contents pane, for NAIP Imagery, click Filter.

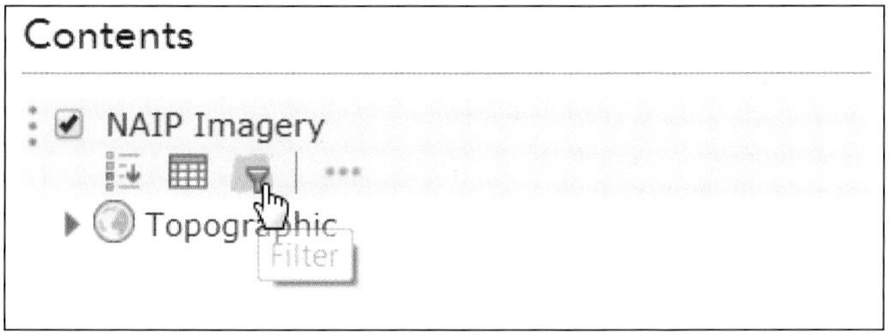

Contents
_____

☑ NAIP Imagery

▶ ◉ Topographic
Filter

14. In the Filter pane, specify AcquisitionDate is May 18, 2012.

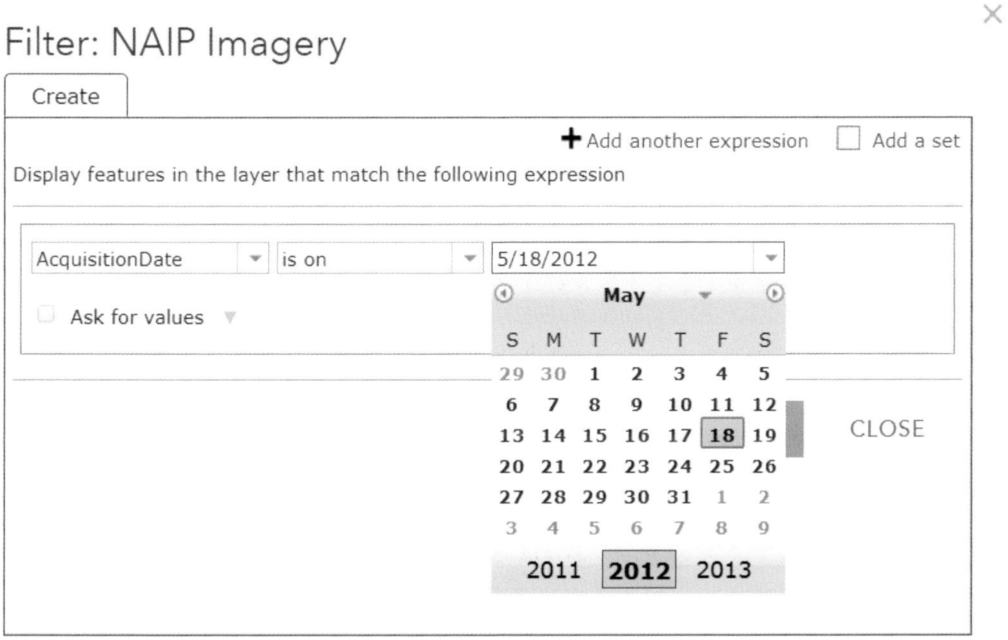

Filter: NAIP Imagery

| Create |

＋ Add another expression   ☐ Add a set

Display features in the layer that match the following expression

| AcquisitionDate | ▼ | is on | ▼ | 5/18/2012 | ▼ |

☐ Ask for values ▼

	◀	**May**	▼	▶

S	M	T	W	T	F	S
29	30	1	2	3	4	5
6	7	8	9	10	11	12
13	14	15	16	17	**18**	19
20	21	22	23	24	25	26
27	28	29	30	31	1	2
3	4	5	6	7	8	9

CLOSE

2011 **2012** 2013

The NAIP image service is published from a mosaic dataset, which manages collections of NAIP images obtained from different dates. By default, the latest images are displayed when you zoom into an area. However, you can use the filter to select the images obtained earlier.

15.   Click Apply Filter.

An image with light color displays.

16.   Click on the image.

From the pop-up window, the image's acquisition date is May 18, 2012.

17.   Click Filter again. Click Remove Filter.

This step removes the filter. The image acquired on May 2, 2016, appears. Next, you will display this image service using different band combinations and stretch types. You will also experiment with different renderers.

18.   Click More Options and click Image Display.

The current renderer is NaturalColor.

19.   Click the drop-down button of Renderer and select User Defined Renderer.

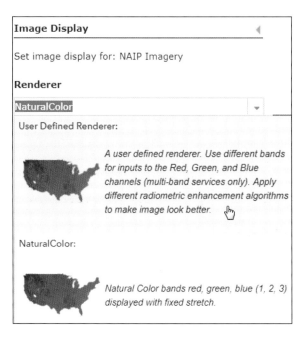

20.   Change the RGB composite to 4, 1, 2, and the stretch type to standard deviation. Leave other settings unchanged. Click Apply.

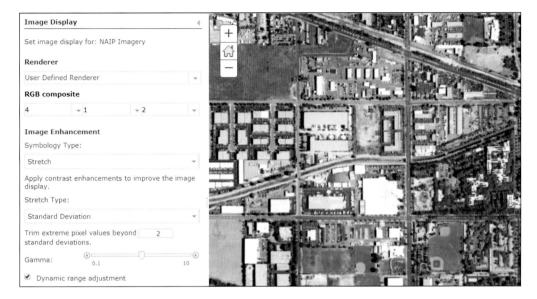

The image service is displayed in false color where the red color highlights vegetation.

21.   Change the Renderer to NDVI_Color. Click Apply.

The image service is rendered using NDVI_Color.

22.   Change the Renderer back to NaturalColor and click Apply. Close the Image Display pane.

23. Click the Save button and select Save from the drop-down menu to save the changes to the map.

You have created a web map using the NAIP image service from the Living Atlas of the World. You will use this web map in the next section to create a web app.

9.2 Create a web app using the Image Interpretation configurable app

In this section, you will use the Image Interpretation configurable app to create a web app that shows the change of vegetation in Redlands, California, between 2012 and 2016.

1. Continuing from the last section, click Home > Home to return to the home page.

2. In the search box, search for **Imagery Interpretation web app template**. The search results show "No items found that meet your criteria. Try clearing some filters to show more items."

3. Clear "Only search in <My Organization Name>".

Content   Groups

Only search in Esri Press -   ◐
Official Book Site

1 - 16 of 33   Sort by: Relevance ∨

Image Interpretation (beta)
🖼 Web Mapping Application by esri_en

Create an app to better visualize and interpret imagery layers in your web map, including tools for change
detection, measurement, recording locations, and more.
Created: Sep 20, 2017   Updated: Dec 6, 2017   View Count: 1280

🌐 ★ ⋯

The Imagery Interpretation application template is found.

4.   Click Image Interpretation.

The Item details pane appears.

5.   On the item details of the Image Interpretation app, select Create a Web
App.

6.   In the Create a New Web App pane, perform the following actions:

- For Title, specify **Vegetation Change in Redlands, CA**.
- For Tags, specify **NAIP**, **NDVI**, and **GTKWebGIS**.
- For Summary, input **Vegetation change based on NDVI**.
- For Save in folder, select Chapter9.
- Click OK.

The Image Interpretation app opens. You are asked to select the web map that will be used in
the app.

7.   In the Select Web Map pane, search NAIP Imagery from My Content.

NAIP Imagery web map is found in the search results.

8.   Select NAIP Imagery and Click Select.

The configuration panel opens.

9. On the General tab, for Title, type **Vegetation Change in Redlands, CA**; for Description, specify **This web app displays vegetation change in Redlands, CA, using Living Atlas's NAIP image service**. Select Enable search tool to add a location search tool in the web app.

10. Navigate to the Imagery Layers tab in the configuration panel and select Enable Layer Selector. Select NAIP Imagery as the Active Layer. Select the Enable Renderer Tool and the Enable Compare. Click the drop-down button under Compare Tools, and select Horizontal Swipe and Vertical Swipe.

11. Navigate to the Image Selector tab and select Enable Image Selector Tool. Click the Display drop-down menu and select Slider and Dropdown List. Select Enable AutoRefresh. Under Imagery Layers, Select NAIP Imagery.

12. Navigate to the Image Date tab. and Select Enable Image Date. For Label, specify **Acquisition Date**. Under Image Layers, select NAIP Imagery. In the expanded attribute list, select AcquisitionDate.

13. Navigate to the Change Detection tab. Select Enable Change Detection Tool. Select Vegetation Index.

14. Click Save.

15. Click Launch to view the app.

The title shows that the current display image was acquired on May 2, 2016.

16. Click Renderer. Click the drop-down button and select NDVI_Color. Click Apply.

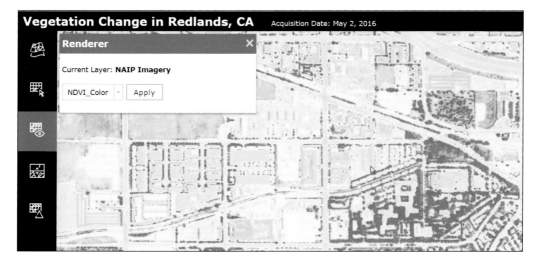

The layer displays with NDVI_Color renderer.

17. Click the drop-down arrow again, change the renderer back to NaturalColor, and click Apply. Close the Renderer window.

18.   Zoom to Esri campus.

19.   Click Image Selector. In the Image Selector window, select Enable Image Selector. Move the slider to the left to 5/18/2012. You can see the image displayed in a lighter color. You will use this image for comparison. Click the arrow to set the image as a comparison layer.

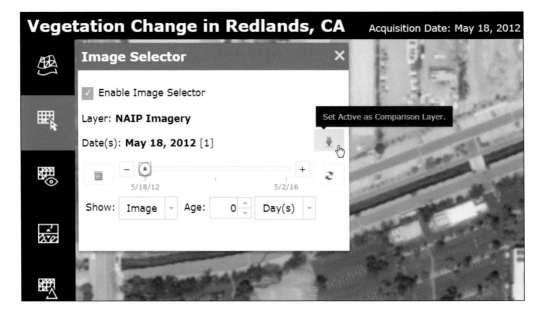

20. Move the slider to the right to 5/2/16. You will use this image to compare with the previously selected image. Close the Image Selector window.

You have selected the two images to compare. Next, you will use the Compare tool and Change Detection tool to compare the images.

21. Click the Change Detection button. In the Change Detection window, leave Method and Mode as their default settings; for Infrared Band, select Band_4, for Red Band, select Band_1. Click Apply. Close the Change Detection window.

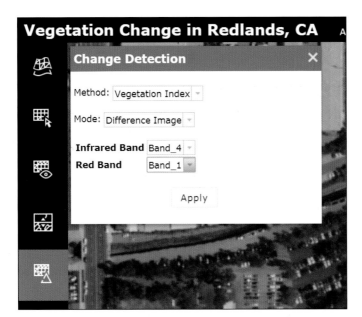

The change detection will calculate the change between two NDVI images. You will observe green and purple colors in the result. Light green represents new vegetation areas in 2016, while purple shows the areas where vegetation disappears in 2016. Next, you will use the Compare tool to perform some comparisons.

22. Click the Compare button. Slide up and down to compare the change detection result with NAIP images. Slide left and right to compare the 2016 with the 2012 NAIP image.

23. Close the Compare window. Close the app.

In this section, you created a web app to show the vegetation change in Redlands, California, from 2012 to 2016.

You used existing image services in your web app. You may want to publish your own image services. The second part of the tutorial will teach you how to publish image services. Please note that you will need ArcGIS Enterprise to publish image services. You will also need image server if you want to publish mosaic datasets as image services.

9

9.3   Explore the multispectral image and create a raster function template file in ArcGIS Pro (optional)

In this section, you will explore a 4-band NAIP multispectral image in ArcGIS Pro. You will examine the image's properties and display the image with different band combinations and stretch types. You will also create pyramids for this raster dataset to improve performance when displayed. Finally, you will create a raster function template file using the NDVI Colorized function to show the vegetation's distribution.

1.   Start ArcGIS Pro.

2.   Add a portal connection.

If you have already added the connection to your portal, skip this step. Otherwise, add a portal connection to either ArcGIS Online or Portal for ArcGIS (reference section 5.1, "Add a Portal Connection in ArcGIS Pro").

3.   Under Create a new project, choose the Map.aptx template.

4.   In the Create a New Project window, for Name, type **Image Service**; for Location, select **C:\EsriPress\GTKWebGIS\Chapter9**; deselect Create a new folder for this project; and click OK.

A new map project containing the default Topographic layer is opened in ArcGIS Pro.

5.   Under the Project tab of the Catalog pane, expand Folders, and then expand folder Chapter9. The image m34117a2sw.tif appears.

6.   Right-click on m34117a2sw.tif and click Properties.

The Raster Dataset Properties pane appears. You can see the general information of this raster dataset, such as number of bands, cell size, spatial reference, and so on. You will notice that the raster dataset's pyramids are absent and its statistics have not been calculated. You will first build pyramids and calculate statistics for this raster dataset.

7.   On the Analysis tab, click Tools to open the Geoprocessing pane.

8. In the search box, type **build pyramids**. Under Search Results, double-click Build Pyramids And Statistics.

Build Pyramids And Statistics appears in the Geoprocessing pane.

9. In the Geoprocessing pane, for Input Data or Workspace, drag m34117a2sw. tif from the Catalog pane. Expand Pyramids Options. For Pyramid resampling technique, select Bilinear. Leave other parameters as their default settings. Click Run.

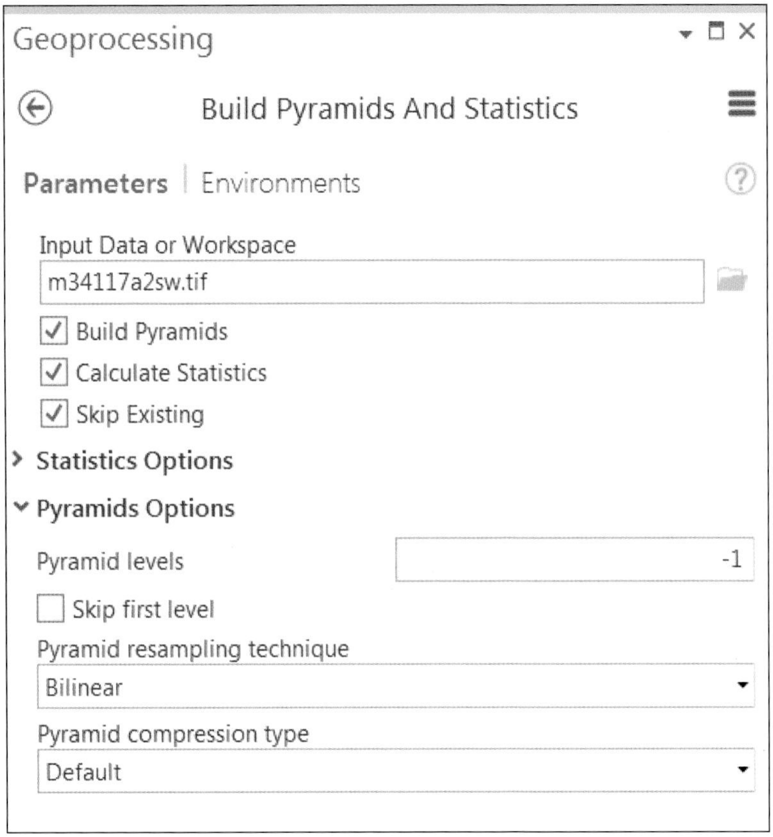

Geoprocessing   ▾ ☐ ✕

  ←   Build Pyramids And Statistics   ☰

  Parameters │ Environments   ⑦

    Input Data or Workspace
    m34117a2sw.tif   📁

    ☑ Build Pyramids
    ☑ Calculate Statistics
    ☑ Skip Existing
  ❯ **Statistics Options**
  ⌄ **Pyramids Options**

    Pyramid levels   -1

    ☐ Skip first level
    Pyramid resampling technique
    Bilinear   ▾

    Pyramid compression type
    Default   ▾

Bilinear interpolation is suitable for displaying images. The resulting image layer adds to the map.

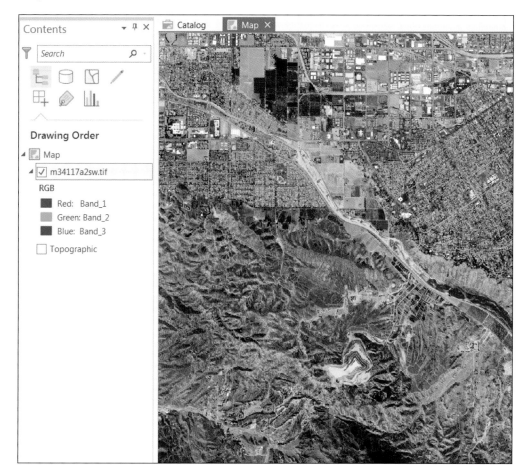

10.   In the Contents pane, right-click m34117a2sw.tif and choose Symbology.

The Symbology pane of the m34117a2sw.tif layer appears. The image currently displays in true color with the band combination of R, G, B = 1, 2, 3 and the default stretch type of Percent Clip. The image appears to be close to what you would expect to see in a normal photograph.

11. In the Symbology pane, change the Band Combination to R, G, B = 4, 1, 2 and the Stretch Type to Standard Deviation.

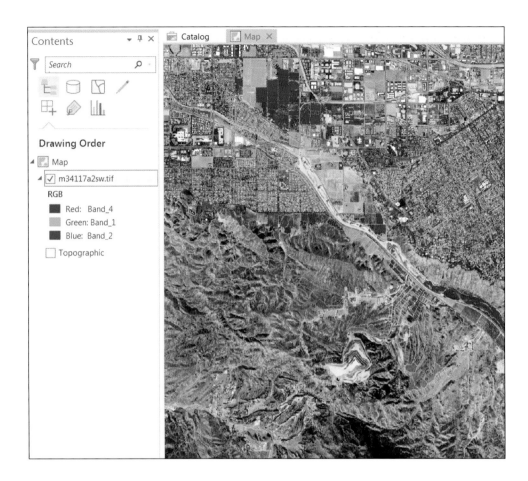

Now, the image is displayed in false color, which highlights vegetation in red.

12. On the upper-left quick access bar, click Save to save your project.

13. On the Analysis tab, click the Function Editor button.

A blank raster function template opens. A raster function template is like a model, which can contain one or more raster functions chained together to produce a processing workflow.

14.   In the Raster Function Editor, click the Raster Functions button 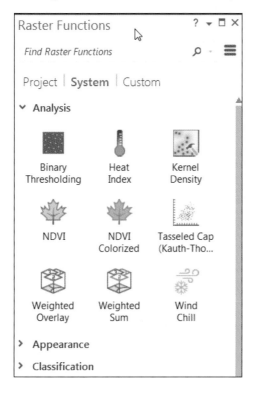 at the top-right corner of the toolbar to open the Raster Functions pane.

15.   In the Raster Functions pane, drag the NDVI Colorized function found in the System tab and drop it into the Raster Function Editor.

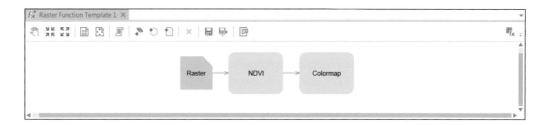

The NDVI Colorized function is a raster function chain, which chains the NDVI and Colormap functions. The NDVI Colorized function uses the NDVI function on the input image to calculate its NDVI values and uses the Colormap function to display the result.

16.   Double-click on the Raster variable to open the Input Dataset window. Browse to Project\Folders\Chapter9 and select m34117a2sw.tif. Click OK.

17.   Double-click the NDVI function to open the Properties pane. On the Parameters tab, for Visible Band ID, select 1; for Infrared Band ID, select 4. Leave Scientific Output unselected. Click OK.

NDVI Properties

General | **Parameters** | Variables

Raster

m34117a2sw.tif

Visible Band ID

1

Infrared Band ID

4

☐ Scientific Output

The NAIP image's band 1 is the red band, and band 4 is the near-infrared band.

18.   Double-click the Colormap function to open its Properties pane. View its settings and leave them as the defaults. Click OK.

19.   On the Raster Function Editor toolbar, click the Save button.

20.   In the Save window, perform the following actions:

- For Name, specify **NAIP_NDVI**.
- For Category, select Project.
- For Sub-Category, select Project1.
- For Description, type **Create a NDVI Colorized function template, which will be used in an image service**.
- Leave other parameters as default.
- Click OK.

In the Raster Function pane, the raster function template that you just saved appears under Project category > Project1 subcategory.

📝 **Note:** Do not specify the raster function template file name using an existing raster function name like NDVI, Clip, and so on. Otherwise, the online raster analysis on the image service may not work.

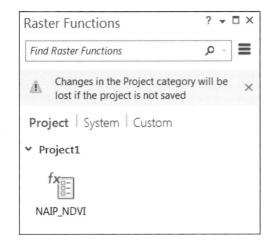

21.   Click the NAIP_NDVI raster function template file to open it.

22.   Click the Create new layer button.

A new image layer is added to the map. This temporary layer is created using the NDVI Colorized function so that you can view the distribution of vegetation. Vegetation appears in green on this layer.

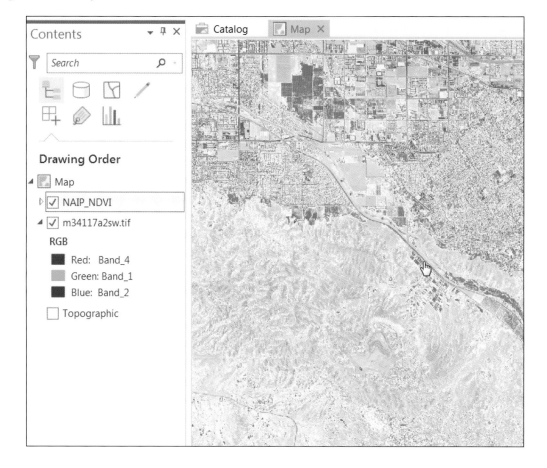

23. Close the Raster Function Editor and Raster Functions pane.

24. Click Save to save the project.

In this section, you created an NDVI Colorized raster function template and applied it to your image. This function template will be used in the image service. If you have Portal for ArcGIS, continue to the next section.

9.4 Share an image service to Portal for ArcGIS (optional)

This section requires Portal for ArcGIS. Make sure you set your Portal for ArcGIS as Active Portal.

1. In the Catalog pane, right-click m34117a2sw.tif and choose Share as Web Layer.

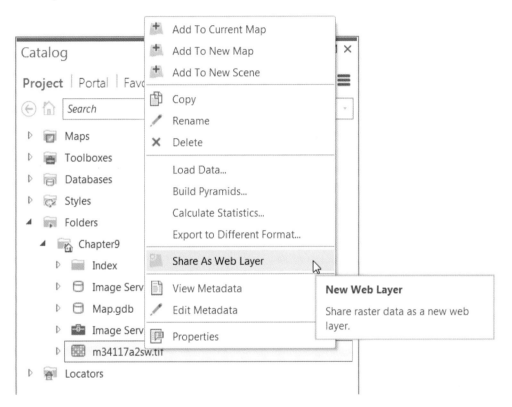

2. If you see a security alert saying the Certificate Issuer for this site is untrusted or unknown, click Yes to proceed.

The alert message is typically because your Portal for ArcGIS uses a self-signed or domain-signed SSL/TSL certificate.

3. In the Share As Web Layer pane, under the General tab, perform the following actions:

   - For Name, specify **Chapter9\NAIP_Redlands**. The web layer will be saved to folder Chapter9 in My Content.
   - For Data, choose Copy all data.
   - For Summary, input **NAIP image service (Redlands, CA) with NDVI Colorized function**.
   - For Tags, specify **Image Service**, **NAIP**, **NDVI**, and **GTKWebGIS**.
   - For Sharing Options, check Everyone.

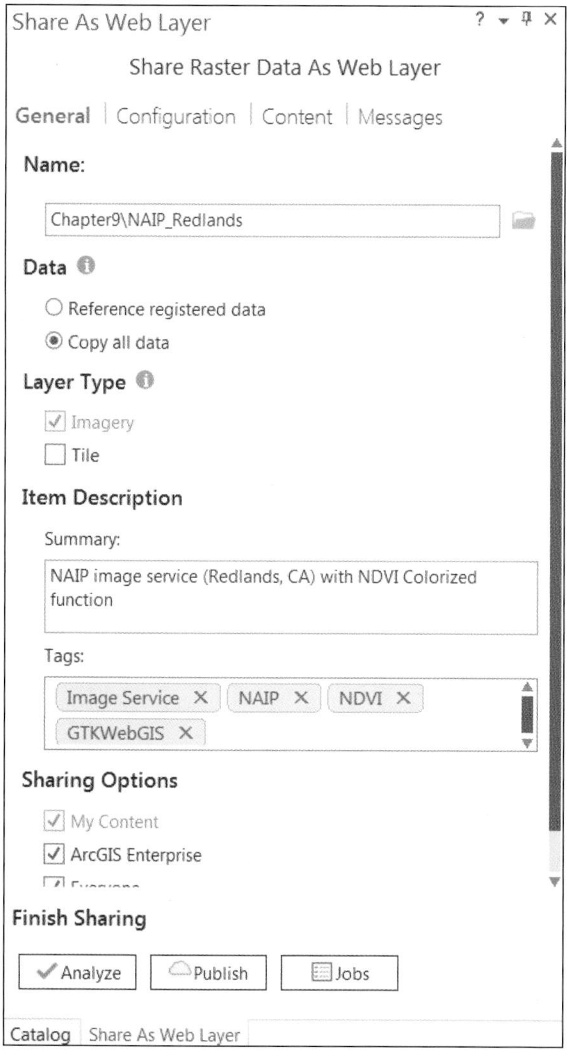
Share As Web Layer
Share Raster Data As Web Layer
General
Configuration
Content
Messages
Name:
Chapter9\NAIP_Redlands
Data
Reference registered data
Copy all data
Layer Type
Imagery
Tile
Item Description
Summary:
NAIP image service (Redlands, CA) with NDVI Colorized function
Tags:
Image Service
NAIP
NDVI
GTKWebGIS
Sharing Options
My Content
ArcGIS Enterprise
Finish Sharing
Analyze
Publish
Jobs
Catalog
Share As Web Layer

Imagery is typically large in file size. You are usually recommended to register the folder containing your imagery in ArcGIS Server Manager, and choose Reference registered data when publishing your image service. Using referenced data will speed up the publishing process. Since the imagery used in this tutorial is not large, you'll choose the Copy all data option.

4. Click on the Configuration tab and click the Configure Web Layer Properties button ✏ to access the configurable properties.

5. In the Image Service Properties pane, expand Raster Functions.

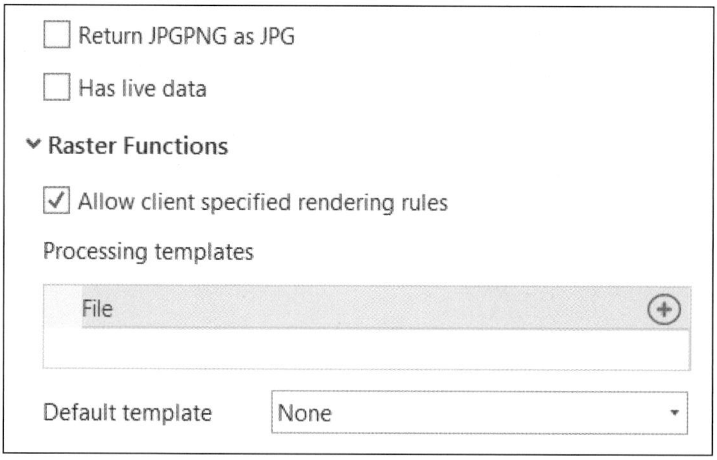

☐ Return JPGPNG as JPG

☐ Has live data

∨ **Raster Functions**

☑ Allow client specified rendering rules

Processing templates

| File | ⊕ |

Default template | None | ▾ |

6. Click the Add processing template button. In the Select raster function template(s) window, browse to **Raster Functions\Project\Project1**, and select NAIP_NDVI.rfx.xml. Click OK.

The raster function template file (.rft.xml) is added under Processing templates.

7. Leave other configuration items as their default settings. Click the Back button ⊙ to return to the Configuration tab.

8. Click the General tab and click the Analyze button.

No errors or warnings are found. Check for any errors or issues. You must resolve all errors before publishing.

9. Once validated, click the Publish button to share your web layer to Portal for ArcGIS.

Once the web layer has successfully published, log into your Portal for ArcGIS to view the image service you just published.

10. Sign in to your Portal for ArcGIS and click Content. In the My Content tab, under Folders, click the Chapter9 folder.

11. Click the NAIP_Redlands More Options button ⋯ and select Add to new map.

The image layer NAIP_Redlands is added in the Contents pane but does not appear in the map.

12. To display the image layer in the map, click More Options and select Zoom to.

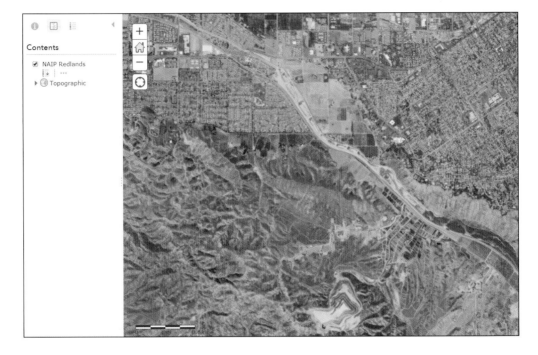

The image layer appears on the map. Next, you will explore the image service.

13. Click More Options and select Configure Pop-up.

By default, Service Pixel Value is set to show in the pop-up.

14. Accept the default setting and select Show Pop-ups. Click OK.

15. Click the image layer.

A Pop-up window shows the image service pixel values of the pixel on which you clicked.

16.   Close the Pop-up window.

17.   Click More Options and select Image Display.

18.   In the Image Display pane, for Renderer, keep User Defined Renderer; for RGB composite, select 4, 1, 2; for Stretch Type, select Standard Deviation; Select Dynamic range adjustment; click Apply.

The image layer is displayed in false color with standard deviation stretch. The vegetation displays in red.

19. In the Image Display pane, for Renderer, select NAIP_NDVI; for Stretch Type, select None. Click Apply.

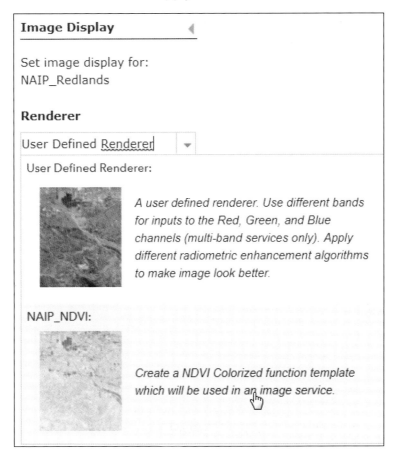

9

A colorized NDVI layer is shown. The vegetation now is displayed in green.

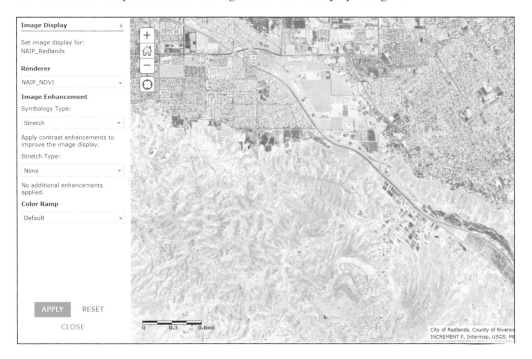

20. To close the Image Display pane, click Close.

21. Click Save > Save. In the Save Map page, for Title, type **NAIP Redlands**; for Tags, specify **NAIP**, **NDVI**, and **GTKWebGIS**; for Summary, type **A web map of NAIP image with NDVI Colorized function**; select Chapter9 to save the web map. Click Save Map.

In this section, you have published your image with the NDVI Colorized function to your Portal for ArcGIS. You examined the image layer and experimented with different band combinations and stretch types. You also applied the NDVI Colorized function to the image service to show the distribution of vegetation.

From this tutorial, you have seen that the image service is not only to be used as a basemap, but also for analysis. Raster function is useful for quick online analysis. For a large collection of raster data, you should use a mosaic dataset to manage the collection and publish the mosaic dataset as an image service.

QUESTIONS AND ANSWERS

1. To publish image services, does my ArcGIS Server need be licensed with an image server role?

   Answer: You can still publish raster datasets or raster dataset layers as image services without an image server license role. However, you will need it to publish mosaic datasets, mosaic dataset layers, or raster dataset layers with mosaic raster function as image services.

2. Can I publish image services to ArcGIS Online?

   Answer: No. Currently, you can only publish image services to ArcGIS Server if you publish from ArcMap or to ArcGIS Enterprise if you publish from ArcGIS Pro.

3. I was trying to publish an image service in ArcGIS Pro. I right-clicked the raster dataset and selected Share As Web Layer. I expected the Share Raster Data As Web Layer pane to open, but instead, the "Sharing selected layer as a web layer" pane opened up. Why?

   Answer: You received the "Sharing selected layer as a web layer" pane because you right-clicked the raster dataset in the Contents pane. The "Sharing selected layer as a web layer" pane is used to publish a map service. Image services are published from the Catalog pane. When you right-click the raster dataset at the Catalog pane, you will receive the Share Raster Data As Web Layer pane, where you can configure the image service.

4. I wanted to publish a cached image service but could not find the setting to set the cache in the Share Raster Data As Web Layer pane. Where can I find the setting?

   Answer: You must copy the data to ArcGIS Server to publish a cached image service. First, choose Copy all data and check Tile as the layer type in the General pane. Then, click the pencil button on the right side of the Tile in the Configuration pane to open the Tile Properties pane, where you can set caching.

5. When publishing an image service, I chose Reference Registered Data. Why did I receive a warning that "24011 Data source is not registered with the server and data will be copied to the server"?

    Answer: This warning indicates that the folder containing the image data has not been registered to the server. Go to ArcGIS Server Manager, under Site > Data Store, click the drop-down arrow of Register and select the Folder option. Specify the folder where your image data is stored and make sure ArcGIS Server has permission to access the folder.

6. I created a web app using the Imagery Interpretation app. When I perform change detection, why does only part of my screen show the change detection result?

Answer: The change detection only detects the changes in the overlapping area of two images. If there is more than one overlapping area within the extent of your screen, only the result of one overlapping area is shown. For this reason, you could only see the change detection result on part of your screen.

  If you want to see the change detection results within a specific area, you should zoom into this specific area and make sure that it is the only overlapping area in your screen's extent. You can click the image and make sure you do not see the image footprint.

7. Where can I download free NAIP 4-band images?

   Answer: You can go to USGS EarthExplorer (**https://earthexplorer.usgs.gov**) to download 4-band NAIP images. If you do not have an account, you will need to register and create a free account to download the products. From the Data Sets tab, under Aerial Imagery, choose NAIP GEOTIFF.

ASSIGNMENTS

Assignment 9A: Create a web app to display high-resolution terrain and perform terrain analysis.

Data: The Terrain image layer from the Living Atlas (focus on the area of Santa Ana, California).

Requirements:

- Display the high-resolution 3-meter FEMA_LIDAR_DTM.
- The Terrain layer can be rendered with Aspect, Slope, and Hillshade functions.
- FEMA_LIDAR_DTM can be compared with other low-resolution DTMs (for example, 30-meter resolution SRTM).

What to submit:

- The URL to your web app

Hint: Use the Image Interpretation web app.

Assignment 9B: Publish an image service and use it in a web app for raster analysis (optional).

Data: C:\EsriPress\GTKWebGIS\Chapter9\DTM_Petaluma.tif, which is the high-resolution DTM (1 feet) of Petaluma, California

Requirements:

- Publish an image service with Aspect, Slope, and Hillshade functions.
- Create a web map based on the image service you published.
- Create a web app using the Elevation Profile template based on this image service.

What to submit:

- The URL to your map service REST endpoint
- The URL to your web map
- The URL to your web app

Resources

"Check out Esri's new Landsat Explorer web app," Emily Windahl, https://blogs.esri.com/esri/
arcgis/2017/03/22/new-landsat-explorer-app (or http://bit.ly/2EszgkQ).

"Depict Land-Use Change with Time-Enabled Apps," https://learn.arcgis.com/en/projects/depict-land
-use-change-with-time-enabled-apps (or http://bit.ly/2Bxmm6z).

"Explore Climate Trends with the Water Balance App," Daniel Siegel, https://blogs.esri.com/esri/
arcgis/2017/09/25/explore-climate-trends-with-the-water-balance-app (or http://bit.ly/2Gdtunq).

"Getting started with the Image Interpretation (beta) configurable app," Beth Romero, https://blogs.esri
.com/esri/arcgis/2017/11/09/getting-to-know-the-new-imagery-interpretation-beta-configurable-app
(or http://bit.ly/2EGKupk).

"Global Elevation Analysis Services Enhances with Higher Resolution Data," Caitlin Scopel, https://blogs
.esri.com/esri/arcgis/2017/04/21/global-elevation-analysis-services-enhanced-with-higher-resolution
-data (or http://bit.ly/2ELF3W6).

"Imagery in Web applications: Should I use a cached map service or an image service?" Sterling Quinn,
https://blogs.esri.com/esri/arcgis/2010/05/04/imagery-in-web-applications-should-i-use-a-cached
-map-service-or-an-image-service (or http://bit.ly/2HigWwx).

"Imagery Sources and Usage in ArcGIS," Miriam Schmids and Nicholas M. Giner, https://www.youtube
.com/watch?v=pnoj24ncZas (or http://bit.ly/2GiKIjB).

"Imagery–Technical: Using Imagery in ArcGIS: An Introduction," Kevin Butler and Gerry Kinn, https://
www.youtube.com/watch?v=zkQkG8HKXlI (or http://bit.ly/2CqR0ve).

"Introducing Esri's World Elevation Services," Rajinder Nagi, https://blogs.esri.com/esri/
arcgis/2014/07/11/introducing-esris-world-elevation-services (or http://bit.ly/2CphNYt).

"Key concepts for image services," http://server.arcgis.com/en/server/latest/publish-services/windows/
key-concepts-for-image-services.htm (or http://bit.ly/2Esyvbu).

"Landsat 8 Imagery Available for Online Users," http://www.esri.com/esri-news/arcnews/
spring14articles/landsat-8-imagery-available-for-online-users (or http://bit.ly/2BxSxmp).

"NAIP Imagery now available as ArcGIS Online Image Layers," ArcGIS Content Team, https://blogs.esri
.com/esri/arcgis/2014/07/02/naip-imagery-now-available-as-arcgis-online-image-layers (or http://
bit.ly/2Es5wZg).

"New and Improved Landsat Apps," jkerski-esristaff, https://community.esri.com/community/education/
blog/2017/04/14/new-and-improved-landsat-apps (or http://bit.ly/2Hjg8r9).

"Tour the World with Landsat Imagery and Raster Functions," https://developers.arcgis.com/python/
sample-notebooks/tour-the-world-with-landsat-imagery-and-raster-functions (or http://bit.ly/
2ByyKmQ).

"Tutorial: Creating a cached image service," http://server.arcgis.com/en/server/latest/get-started/
windows/tutorial-creating-a-cached-image-service.htm (or http://bit.ly/2C20x00).

"Use the Image Interpretation app template to create an imagery app in minutes," Emily Windahl,
https://blogs.esri.com/esri/arcgis/2017/10/16/image-interpretation-app-template (or http://bit.
ly/2EuISiV).

CHAPTER 10

Web GIS programming with ArcGIS API for JavaScript

Programming is required for enhancing existing software products, creating new ones, and transforming the creative ideas in your mind into realities. In previous chapters, you learned to create web apps using various configurable app templates and app builders. These apps provide tremendous functionality, but they may not meet all your project requirements. In such cases, you must program your own apps or customize existing web apps. Web GIS programming can be accomplished at the client/browser side, web server side, and the database side, depending on your actual needs. This chapter focuses on browser-side programing with JavaScript, the most popular programming language on the web today. Built on top of JavaScript, ArcGIS API for JavaScript provides libraries for you to add GIS capabilities in JavaScript apps. In this chapter, you will learn the basics of JavaScript API, the essential skills to debug JavaScript, and the general workflows to adapt JavaScript samples, handle events, and incorporate widgets and tasks.

Learning objectives

- *Understand the basics of JavaScript API.*
- *Debug JavaScript.*
- *Adapt JavaScript samples.*
- *Develop 2D and 3D GIS apps using JavaScript.*
- *Handle JavaScript events.*
- *Use widgets.*
- *Use tasks.*

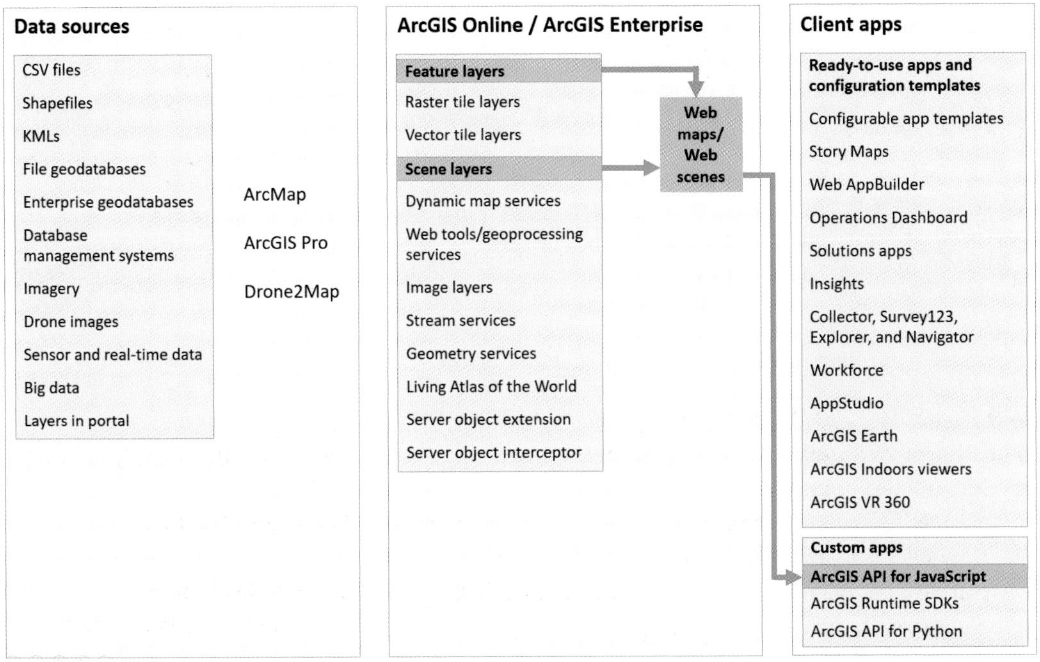

Data sources	ArcGIS Online / ArcGIS Enterprise	Client apps
CSV files	**Feature layers**	**Ready-to-use apps and configuration templates**
Shapefiles	Raster tile layers	Configurable app templates
KMLs	Vector tile layers	Story Maps
File geodatabases	**Scene layers**	Web AppBuilder
Enterprise geodatabases ArcMap	Dynamic map services	Operations Dashboard
Database management systems ArcGIS Pro	Web tools/geoprocessing services	Solutions apps
Imagery Drone2Map	Image layers	Insights
Drone images	Stream services	Collector, Survey123, Explorer, and Navigator
Sensor and real-time data	Geometry services	Workforce
Big data	Living Atlas of the World	AppStudio
Layers in portal	Server object extension	ArcGIS Earth
	Server object interceptor	ArcGIS Indoors viewers
		ArcGIS VR 360

Web maps/ Web scenes

Custom apps
ArcGIS API for JavaScript
ArcGIS Runtime SDKs
ArcGIS API for Python

ArcGIS offers many ways to build web apps. The green lines in the figure highlight the technology presented in this chapter.

Web GIS programming overview

Depending on the requirements of your Web GIS products and projects, you can implement the requirements at the client side, web server side, or database side.

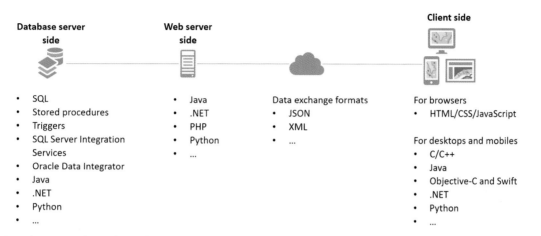

Main languages for Web GIS development at the client, web server, or database side.

Client-side programming

Client-side programming creates apps that can run inside web browsers or run natively on desktops and mobile devices.

- Browser-based app programing primarily uses JavaScript, HTML, and CSS today. Plug-in technologies have been fading away.
- Native desktop and mobile apps require specific languages for specific platforms.

Developers can use these general-purpose programming languages and add GIS capabilities using ArcGIS Web APIs and SDKs. A variety of APIs and SDKs target different use cases and platforms. Developers can choose the right API or SDK based on the target platforms, developer team skills, project functional requirements, and performance requirements.

ArcGIS provides **a suite of APIs and SDKs that developers can use to develop browser, desktop, and mobile native apps.**

	Android	iOS	Windows	mac OS	Linux
ArcGIS Runtime SDK for Android	✓				
ArcGIS Runtime SDK for iOS		✓			
ArcGIS Runtime SDK for Java			✓	✓	✓
ArcGIS Runtime SDK for macOS				✓	
ArcGIS Runtime SDK for .NET	✓	✓	✓		
ArcGIS Runtime SDK for Qt	✓	✓	✓	✓	✓
ArcGIS API for JavaScript	✓	✓	✓	✓	✓
Web AppBuilder for ArcGIS	✓	✓	✓	✓	✓
AppStudio for ArcGIS	✓	✓	✓	✓	✓
ArcGIS API for Python *	✓	✓	✓	✓	✓

* ArcGIS API for Python can run inside web browsers and display results in web browsers, including mobile browsers and desktop browsers.

Web server-side programming

Web servers accept web requests from clients, for example through HTTP or web socket protocols. Web servers perform server-side operations and then serve clients with responses. Web servers are also containers that allow certain scripts such as Java and .NET to run inside the containers. Such programs allow web developers to implement business logics on the web server side. There are mainly two types of web application servers:

- Microsoft IIS (Internet Information Services): IIS supports .NET natively and can support other languages through add-ons.
- Other than IIS, most web application servers are Java-based. This category includes Apache, Tomcat, GlassFish, and WGS (Google Web Server). They support Java natively and can support other languages through add-ons.

Developers can use general-purpose programming languages such as Java, .Net, and Python for server-side programming and add GIS capabilities using server-side GIS libraries. For example, here are some ways for server-side GIS programing.

- **Server object extensions (SOEs):** SOEs allow developers to use .Net or Java to create new web service operations to extend the base functionality of map, image, and feature services. SOEs are appropriate if you have some well-defined business logics to perform, but they are not easily accomplished using ArcGIS client APIs. SOEs can be implemented using ArcObjects, which is one of the core components of ArcGIS and allows the most flexibility in writing GIS functions.
- **Server object interceptors (SOIs):** Allow developers to use .Net or Java to intercept requests for existing, built-in operations of map, image, and feature services. This allows developers to execute custom logic and alter the behavior of these services by overriding

existing operations in a way that is seamless for existing clients. For example, SOIs can be used to filter access to specific layers or even filter data within a layer based on the user's role. SOIs can be written using ArcObjects.

- **ArcPy:** ArcPy is a Python site package that provides a useful and productive way to perform geographic data analysis, data conversion, data management, and map automation with Python. Python code with ArcPy can be published as geoprocessing services or web tools to ArcGIS Enterprise.

Database server-side programming

Programming on the database side eliminates the need to transfer data over the network. This is useful when the data size is large, especially in the case of big data.

Database-side programing technologies include Structured Query Language (SQL), stored procedures, and triggers. Some databases can run Java code inside the database directly. Databases can also be accessed by Java, .NET, and Python via database connectors such as Open Database Connectivity (ODBC), Java Database Connectivity (JDBC), and Python ODBC (pyodbc).

Data exchange format

Contemporary web technology prefers to separate the data from the formatting and the exchanging of data from the presentation. The data format used for exchanging messages between the server and the client can influence the load to bandwidth and therefore the efficiency of the application. The main formats for exchanging GIS data over the web are JSON and XML. JSON is much more widely used than XML today in Web GIS programming.

JavaScript, HTML, and CSS

This chapter focuses on JavaScript, the mostly widely used programming language today. Almost all web pages today include some JavaScript code. JavaScript is a bridge between web browsers and servers. JavaScript interacts with servers to use the servers' capabilities and works with web browsers to make web pages dynamic and interactive.

Unlike web services and other server-side programs, which run on the server side, JavaScript typically runs on the browser side. JavaScript has more cross-platform capabilities than native desktop apps and mobile apps. All desktop and mobile web browsers support JavaScript. With responsive web design approaches, a single JavaScript web app can run on both desktop browsers and mobile browsers, and across a wide range of screen sizes.

JavaScript is easy to start using. You do not have to use a professional integrated development environment (IDE), and you do not have to compile your JavaScript code. You can write JavaScript using any clear text editor and then load and run the source code in a web browser.

JavaScript works in conjunction with HTML and CSS for building web apps. To develop web apps using JavaScript, you must know at least the basics of HTML and CSS. You can quickly learn the basics of HTML, CSS, and JavaScript at the tutorial website, **http://www.w3schools.com**.

- HTML is a markup language used to package your content.
- CSS is a formatting language used to style content.
- JavaScript creates dynamic and interactive features for your web pages.

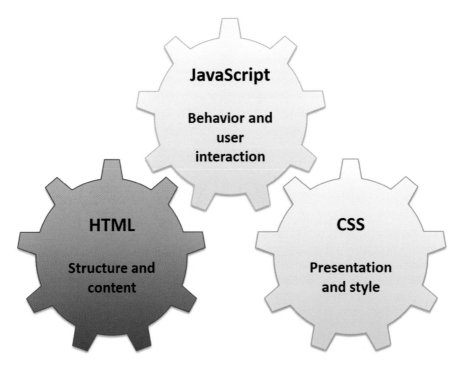

HTML, CSS, and JavaScript work together in web app development.

Most developers write JavaScript apps on top of JavaScript frameworks, such as Dojo, JQuery, Bootstrap, Angular JS, and React. These frameworks provide commonly used components or widgets, help reduce development efforts, and enhance cross-browser capability. JavaScript API is based on the Dojo framework, but developers can use it with other frameworks.

HTML5

HTML5 is the fifth and current major version of HTML standard. Although not technically part of HTML5, JavaScript and CSS3 (CSS standard version 3) are often considered parts of HTML5. HTML5 comes with many new features that are of great value in Web GIS.

- Canvas and WebGL provide interfaces for displaying 2D and 3D data. WebGL allows browsers to display 3D maps without plug-ins. WebGL enables ArcGIS high-performance feature layers. ArcGIS high-performance feature layers are optimized by ArcGIS Online and ArcGIS Enterprise and are rendered by WebGL on the browser side.
- Geolocation can locate and track mobile devices and thus allows browser-based mobile data collection and location-based services.
- WebSocket allows servers to push data to the client side in real time, which is an important foundation for real-time GIS.
- Local Storage and Web SQL support offline use of Web GIS, including offline data editing.
- Video and audio tags make it easier than ever before to add video and audio media to Web GIS apps.
- Input-type attributes make it easy to create user-friendly forms for attribute data collection and for mobile apps.

JSON

JavaScript API relies heavily on JSON for storing and exchanging data. JSON is a lightweight computer data interchange format. It is derived from the object literals of JavaScript, as defined in the European Computer Manufacturers Association (ECMA) Script Programming Language Standard. JSON is smaller in size than XML, and it is more efficient to parse in JavaScript where JSON is a native data type. JSON includes the following syntax rules:

- Data is in field_name:value pairs.
- Data is separated by commas.
- Curly braces hold objects.
- Square brackets hold arrays.

For example, the following JSON code describes an array of two students. Each student is represented with a name field and name value, a hobby field and hobby value, and an address field and address value. Each address value—with a street field and street value and a zip field and zip value—is another JSON object.

```
{
    "students": [{
            "name": "John",
            "hobby": "Basket ball",
            "address": {
                    "street": "380 New York St",
                    "zip": 92373
            }
    }, {
            "name": " Lisa",
            "hobby": " Movie",
            "address": {
                    "street": "270 State Ave",
                    "zip": 92000
            }
    }]
}
```

ArcGIS REST API

JavaScript API relies on the ArcGIS REST API to utilize the capabilities on the GIS server side. In previous chapters, you learned how to publish various web services or web layers and tools using ArcGIS Online and ArcGIS Enterprise. These services and web maps expose their capabilities via ArcGIS REST API. ArcGIS JavaScript APIs interact with the servers via the REST API. In terms of server-side capabilities, the JavaScript API can deliver no more than what the REST API can deliver.

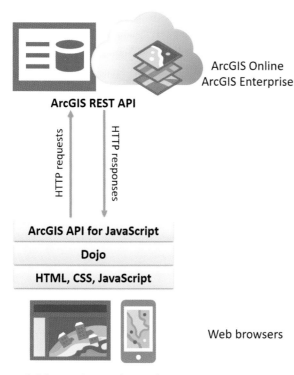

ArcGIS for JavaScript relies on the ArcGIS REST API to interact with ArcGIS Online and ArcGIS Enterprise.

With the ArcGIS REST API, all resources and operations are exposed through a hierarchy of endpoints or URLs. The REST endpoint of a web service on ArcGIS Online is typically in the following format:

http(s)://<server>/<orgId>/arcgis/rest/services/<serviceName>/<serviceType>

For example, the following is the URL of a hosted feature service:

https://services2.arcgis.com/No7KRrFgpO516cMP/arcgis/rest/services/ NaturalDisasters/FeatureServer

If you want to try this URL, you can type **http://arcg.is/2hOnyHC**, short for the long URL, in your web browser address bar. The REST URL of a layer in the service is the service URL appended with a layer number, starting with 0 for the first layer. For example, the following URL is the URL of the second layer in the feature service just mentioned.

https://services2.arcgis.com/No7KRrFgpO516cMP/arcgis/rest/services/ NaturalDisasters/FeatureServer/1.

If you want to try this URL, you can type **http://arcg.is/2zulZts**, short for the long URL, in your web browser address bar.

These REST URLs are what JavaScript API uses to utilize these server-side resources. For example, to perform a query on the feature layer just mentioned, with a where clause of MAGNITUDE>5, returning geometries, and retuning the LOCATION, MAGNITUDE, and YEAR attribute fields, in pretty JSON (easy-to-read JSON format), the URL is

https://services2.arcgis.com/No7KRrFgpO516cMP/arcgis/rest/services/ NaturalDisasters/FeatureServer/1/query?where=MAGNITUDE%3E5&outFields=LOCATIO N%2CMAGNITUDE%2CYEAR&returnGeometry=true&f=pjson

If you want to try this URL, type **http://arcg.is/2mUYd47**, short for the long URL, in your web browser address bar, and then review the URL request and the JSON response.

Building these URLs manually is complicated and error prone. When you use JavaScript API, the API builds these URLs for you automatically.

JavaScript API

JavaScript API adds GIS capabilities to JavaScript apps. The API has two main categories for its GIS capabilities:

- Interact with GIS servers, including ArcGIS Online and ArcGIS Enterprise, via ArcGIS REST API to provide mapping, querying, editing, analysis, and other GIS functions.
- Interact with users and display the responses of GIS servers in maps, views, pop-ups, charts, and other formats in web browsers.

The JavaScript API is available to developers in several ways. The most common way is to use the Content Delivery Network (CDN) hosted version. This way, you do not have to download and install the API. You simply reference the API using the <script> and <link> tags, as you will see in the tutorial section. In some cases, you may need to use a locally hosted version of the JavaScript API, for example, if you are in a restricted network environment with no internet access. You can download and install the API on your local server.

As a developer, you will frequently use the JavaScript API website (**http://developers.arcgis .com/javascript**), on which you can find guides, API references, samples, and the link to the JavaScript API forum on GeoNet (**http://geonet.esri.com**), a place where you can share, chat, and collaborate with other ArcGIS users.

The Help document site for JavaScript API. The site may look different from what you see here.

JavaScript API main classes

JavaScript API provides many classes for you to create objects. This list includes the following most commonly used classes:

- **View:** Includes the MapView class and the SceneView class.
  - **MapView:** Displays maps with 2D renderers. A MapView has properties including extent, center, rotation, scale, and more.
  - **SceneView:** Displays 3D scenes and maps with 3D renderers. SceneView properties include center, center scale, camera, and more.
- **Map:** Manages layers that can be added and removed from a web map dynamically or can be loaded from an existing web map.
- **WebScene:** Manages scene layers that can be added to and removed from a WebScene dynamically. A WebScene can also be loaded from an existing web scene.
- **Layer:** The fundamental component of maps and web scenes. This class includes subclasses representing the feature layer, graphics layer, tiled layer, stream layer, vector tile layer, web tiled layer, map image layer, image layer, elevation layer, and scene layer.
- **Render:** Defines how to visually represent each feature in a layer. This class includes such subclasses as simple, class break, and unique value renderers.
- **Symbols:** Displays points, lines, polygons, and text in 2D. This class includes many subclasses such as SimpleMarkerSymbol, SimpleLineSymbol, SimpleFillSymbol, and more.
- **3D Symbols:** Displays points, lines, polygons, and text in 3D. This class includes many subclasses such as IconSymbol3DLayer, LineSymbol3DLayer, ExtrudeSymbol3DLayers, and more.
- **Task:** Includes subclasses such as QueryTask, Geoprocessor, and RouteTask. These subclasses allow you to perform attribute or spatial searches on individual layers or all layers in a map service, use geoprocessing tasks, find optimal routes, and more.
- **Portal:** Provides a way to build apps that work with content from ArcGIS Online and Portal for ArcGIS. For example, you can load an ArcGIS Online web map or web scene, with the layers already configured with renderers, symbols, and pop-ups.

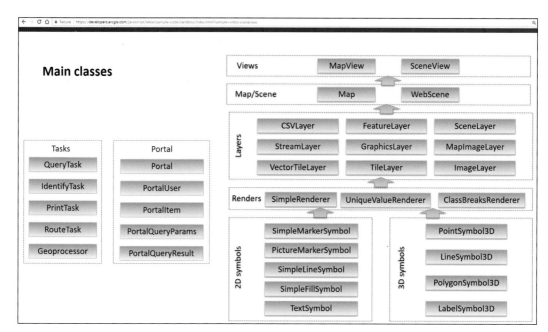

Main classes in JavaScript API

Class properties, methods, and events

Each JavaScript API class includes certain properties, methods, and events:

- **Properties:** A MapView, for example, includes extent, center, rotation, scale, zoom, and other properties. You can obtain and set the properties of an object by using this simple notation: `objectName.propertyName`. To obtain and set the zoom level of a map view, for example, you can use `mapView.zoom`.

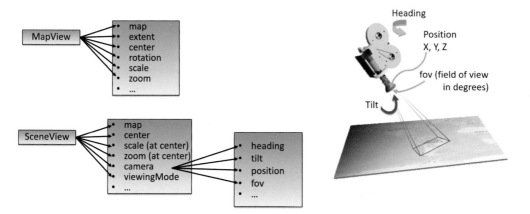

Classes, for example, MapView and SceneView, in JavaScript API have properties.

- **Methods:** Actions or functions that a class can perform. You can call or run the method in this format: `objectName.methodName(parameters)`. For example, you can add a layer to a map using the add method of a map: `map.add(layer)`.
- **Events:** Events happen to the elements in your app, such as when an object is ready, started, changed, completed, moved, displayed, or has an error. You may trigger some functions in response to these events.
  - To monitor an object's property change, use the `.watch(property, callback)` method. For example, you can watch and handle a mapView's extent property change event with the following code:

```
mapView.watch("extent", function(response) {
console.log("the response object is the new extent");
});
```

  - To handle the result of a task or class, use a promise. Promises play an important role in the JavaScript API. In essence, a promise is a value that "promises" to be returned whenever the process completes. Promises are commonly used with the `.then(callback, errback)` method. For example, you can handle a queryTask's result with the following code:

```
queryTask.execute(parameters).then(
 function getResults(queryResult) {},
 function getErrors(err) {}
 );
```

Develop apps using JavaScript API

You will generally follow six steps when you use JavaScript API to write a web app:
1. Reference the JavaScript API.
2. Load the API modules needed for your functions.
3. Create your map or scene.
4. Create your 2D map view or 3D scene view.
5. Define the page content, especially the space(s) to hold your map and scene views (divs).
6. Style the page.

When you adapt an existing sample or combine multiple samples, you will not use all the steps because some of the steps are already done in the samples. You can explore the samples to learn JavaScript API quickly and easily. Most samples on the API website have an Explore in the sandbox button. You can click the button, modify the sample source code in a text box in your web browser, and run your code in the browser without installing it on your own web server.

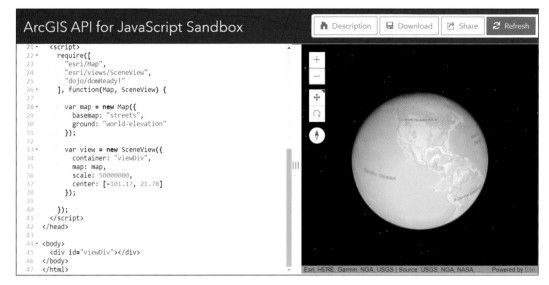

You can learn to use and explore code samples directly in your browser.

Adapting samples and combining samples

Using samples to start your web app development project is cost effective. There are plenty of samples on the JavaScript API website, and you can start with these samples.

In general, adapting a JavaScript sample for your own requirements involves three important steps:

1. Replace the web service or layer URL(s) with your service or layer URL(s).
2. Replace the attribute field names with names for the new service(s) or layer(s) and the related values of the fields.
3. Replace the related symbols, which often involve feature layers and graphics. For example, a sample app highlights some lines using a line symbol. If you want this sample to work with a point layer, match the feature type by changing the line symbol to a point symbol.

Combining multiple samples can be more involved, but in general, you must load the required modules that each sample needs. If possible, you should keep the source code from different samples as separate functions or classes. This practice can help you avoid code conflicts and more easily maintain your code.

Incorporating widgets

A widget is a self-contained component that you can easily incorporate into your JavaScript apps. JavaScript API provides many widgets, such as BasemapGallery, BasemapToggle, Compass, Expand, LayerList, Legend, Locate, NavigationToggle, Popup, Print, ScaleBar, Search, and Track widgets. Typically, you can incorporate widgets into your web apps in three steps.

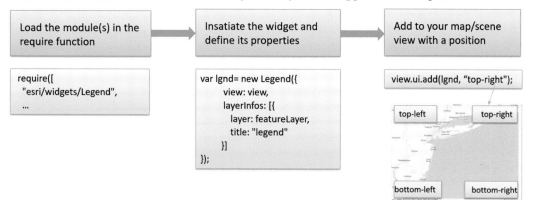

Typical steps to add a widget to your web app.

> **Note:** The widgets in JavaScript API are different from the widgets in Web AppBuilder. The former can be directly used in the latter, but the latter can't be directly used in the former.

Add layers to your apps

You can add layers to your apps in the following two ways:

- **Using web maps and scenes:** Web maps and scenes have already integrated the layers needed, stylized these layers, configured their pop-ups, labels, visible ranges, and configured who can access the maps or scenes. JavaScript API can load web maps and scenes into your app using a few lines of code, and preserve the layer configurations and security settings.
- **Adding layers manually:** This approach gives you the flexibility to add layers on the fly as needed, but this approach is much more difficult than the web map/scene approach. In addition to writing your own code to add the layers, you often need to write more code to configure the layer styles, pop-ups, labels, and visible scale ranges. You must first choose the appropriate layer types to use. For example, you use FeatureLayer for hosted feature layers and sublayers in map image layers, TileLayer for tiled map services, VectorTileLayer for vector tiles, and StreamLayer for data streamed using HTML5 WebSockets. As ArcGIS rolls out high-performance feature layers, MapImageLayer type will be used much less, and FeatureLayer will be used much more for operational layers.

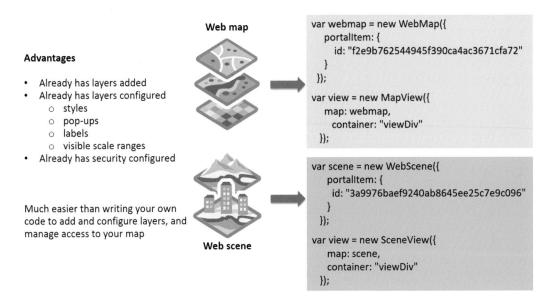

It's much easier to load layers using web maps and web apps than writing your own code to add layers and configure them.

More advanced JavaScript samples

You can start with the samples on the JavaScript API website. As your programing skills and needs grow, you can create custom apps based on more advanced apps or technologies:

- **ArcGIS Online and Portal for ArcGIS configurable apps:** The source code for these apps is hosted on Esri GitHub (**https://github.com/esri**) along with detailed help documents. You can download the source code and customize the apps.
- **ArcGIS solutions apps:** You can explore the web apps at **http://solutions.arcgis.com/ gallery**, select an app you like, download its source code, and enhance the apps.
- **Web AppBuilder for ArcGIS Developer Edition:** This product provides a complete suite of widgets and versatile themes. You can customize and extend it by creating your own widgets and themes (**https://developers.arcgis.com/web-appbuilder**).

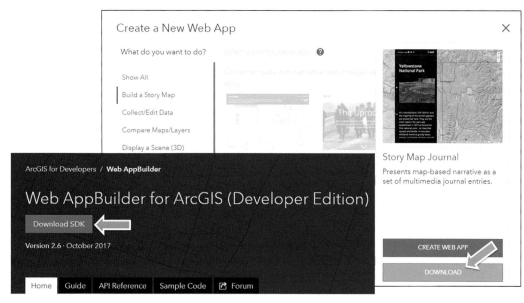

You can create JavaScript apps by extending ArcGIS configurable apps and Web AppBuilder.

10

IDE and JavaScript debugging

Inevitably, you will make typos or other errors while writing programs. JavaScript IDEs such as WebStorm, Sublime Text, and Microsoft Visual Studio can ease your programming experience by providing syntax highlighting and IntelliSense. IntelliSense, a context-aware, code-completion feature, reduces misspellings, typos, and other common mistakes.

As a developer, you often must use debugging tools. Otherwise, even identifying a small typo can be frustrating. Most web browsers provide developer tools. These tools have the following capabilities:

- Display JavaScript errors that arise in the consoles at runtime.
- Pause the execution of your code at the breakpoints you set so that you can examine the variable values and the status of the document object model (DOM).
- Monitor network activities so that you can analyze the network usage and improve the performance of your code.
- Inspect HTML elements and modify their styles and layouts.

This tutorial

In this tutorial, you will create a web app to display earthquakes and tectonic boundaries in 2D and 3D views. You will link two views so that the 3D view will follow the 2D view as you pan, zoom, and rotate the 2D view. You will add a widget and the query capability to your app.

Data: You are provided with a web map and a web scene.

System requirements:

- A web browser for viewing your apps and learning debugging. This tutorial was written based on Chrome version 64 and tested on Chrome 66, but you can use any web browser that allows you to view the tutorial HTMLs using the file:/// URL. Your browser must support WebGL to display 3D scene views. You can find detailed system requirements by visiting the ArcGIS Scene Viewer requirements page.

- Notepad++, a pure text editor, or a JavaScript development environment for editing JavaScript code. You can read your JavaScript code more easily in Notepad++ than in Notepad. You can download Notepad++ at **http://www.notepad-plus-plus.org**. By using this application, you will not need to learn a professional IDE for this chapter.

10.1 Understand the basics of 2D views and 3D views

In this section, you will review two basic samples on 2D and 3D views on the JavaScript API website.

1.   Open a web browser, go to JavaScript API website (**https://developers .arcgis.com/javascript**).

2.   Click the Sample Code tab.

3.   Under Get Started, click Intro to MapView (2D).

4.   On the right side of the screen, scroll down to read the step-by-step instructions and become familiar with the general steps for developing JavaScript apps.

☐ **Note:** Try to understand the source code, but you do not need to understand every single line of the code.

5. Scroll up and click Explore in the sandbox.

6. On the left, read the source code, and on the right, navigate the 2D view.

   - Left-click and drag to pan.
   - Right-click and drag to rotate.
   - Scroll forward or backward to zoom in or out.

7. In the source code, find basemap, and change its value from streets to **osm**. On the right side, click Refresh.

The basemap changes to OpenStreetMap.

8. In the source code, change the value for zoom to **3**, and change the value for center to **[-110, 45]**. On the right side, click Refresh.

The map extent changes to cover most of the US. Next, you will explore a 3D map view sample in the similar way you did with the 2D sample.

9. To return to the Sample Code page, click your web browser's Back button.

10. Under Get Started, click Intro to SceneView (3D).

11. On the right side of the screen, scroll down to read the step-by-step instructions.

The instructions are similar to those of the 2D map view sample.

12. Scroll up and click Explore in the sandbox.

13. On the left, read the source code, and on the right, navigate the 3D view.

    - Left-click and drag to pan.
    - Right-click and drag to tilt.
    - Scroll forward or backward to zoom in or out.

14. In the source code, change the value for basemap to hybrid. On the right side, click Refresh.

The basemap changes to the hybrid basemap.

15.   In the source code, change the value for center to [-110, 45]. On the right side, click Refresh.

The map centers to the area of the US. If you want to save the code in the sample, you can click Download.

10.2 Load web maps and web scenes

Loading layers via web maps and web scenes is much easier than loading individual layers and configuring them in JavaScript manually.

1.   Open a web browser, go to **http://arcg.is/2fd7ZYK**.

This URL goes to the Load a basic web map sample. You can also find the sample by searching **Load webmap**.

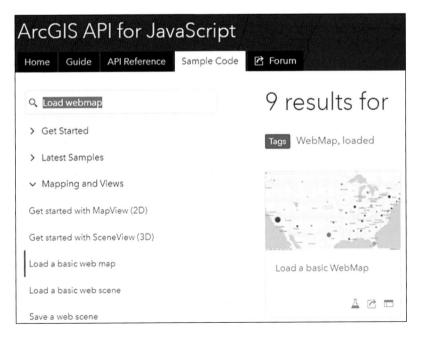

2.   In the sandbox, click Explore.

3.   Read the source code and note the following steps:

- The app loaded the MapView and WebMap modules.
- The app created a web map based on a web map id.
- The app created a map view, which displays the web map in a div.

4.   In the source code, change the id to **08656515afaf4d0587d4f99b9909ddfc** (you can copy the id from **C:\EsriPress\GTKWebGIS\Chapter10\load_basic_ webmap_key.html**).

The updated line of code should look like the following line:
```
id: "08656515afaf4d0587d4f99b9909ddfc"
```

5.   Click Refresh. Note the web map displays on the right side.

The new web map includes two layers, an Earthquakes layer and a Tectonic Plates boundary layer. Next, you will load a 3D web scene.

6.   Click the browser's Back button to go back to the JavaScript Samples page, search for Load web scene, click the Load a basic web scene sample.

You can also go to **http://arcg.is/2gjpK8V** directly.

7.   In the sandbox, click Explore.

8.   Read the source code and note the following steps:

- The app loaded the SceneView and WebScene modules.
- The app created a web scene based on a web scene id.
- The app created a scene view, which displays the web scene in a div.

9.   In the source code, change the id to **ae2631226f9b4883942a1d2423e29772** (you can copy the id from **C:\WebGISData\Chapter14\load_basic_ webscene_key.html**).

10.   Click Refresh. Note the web scene displays on the right side.

The web scene includes two layers, an Earthquakes layer and a Tectonic Plates boundary layer. This section demonstrated that it's simple to load layers by loading web maps or web scenes.

10.3 Debug JavaScript and monitor HTTP traffic

You can easily work with the sandbox source editor in the beginning, but the editor has some limitations. If you refresh the page or leave the page, you will lose all your work. In addition, the editor is not convenient for debugging your code.

In this section, you will edit JavaScript files locally and learn how to debug JavaScript. Learning this skill can save you a lot of frustration and greatly increase your productivity when you develop apps.

📄 **Note:** Typically, you should deploy your JavaScript code to a web server, load your apps in a web browser via HTTP or HTTPS protocol, and then debug your code. To keep the tutorial short and easy, you will load your file into a web browser directly without deploying the file. See the "Questions and answers" section for related information.

1.   Open your web browser, navigate to C:\EsriPress\GTKWebGIS\Chapter10\ webmap_debug.html.

The app displays the same web map you saw in the previous section. Next, you will introduce intentional errors and debug your code using the developer tools.

2.   Start Notepad++ or another JavaScript editor, and open C:\EsriPress\ GTKWebGIS\Chapter10\webmap_debug.html.

3.   In the source code, find new `MapView` and change it to new `mapView` (change the **M** to lowercase). Save your code.

4.   In your web browser, reload C:\EsriPress\GTKWebGIS\Chapter10\webmap_ debug.html. Notice that nothing is displayed.

Next, you will use your web browser to find out where the code is wrong.

5.   With your browser window as your active window, press the F12 key to open the Developer Tools.

Alternatively, you can open the Developer Tools by clicking the Options button and clicking More tools > Developer tools.

6. In the Developer Tools window, click Console in the top menu bar. Notice the error message, along with the line that contains the error. If you don't see the error message, reload the page in your web browser.

7. Click the file name and the line number of the error.

The source code appears with the error line highlighted, so that you can more easily find the problem.

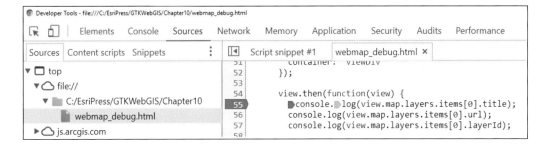

8. In Notepad++, locate the line with the error. Change new mapView back to new MapView (change the **m** back to uppercase **M**) and save your program.

9. Reload your page in your web browser. Notice that the scene appears in the browser, and no error messages appear in the console.

Next, you will learn how to set breakpoints.

10. In the Developer Tools menu bar, click Sources, click webmap_debug.html in the folder tree to open the file. Click line number 55.

```
  Developer Tools - file:///C:/EsriPress/GTKWebGIS/Chapter10/webmap_debug.html

  ⊡ ⟦⟧ |   Elements   Console   Sources   Network   Memory   Application   Security   Audits   Performance

  Sources  Content scripts  Snippets     ⋮    ◧   Script snippet #1    webmap_debug.html ✕
▼ ☐ top                                     51          container:   viewDiv
  ▼ ⟲ file://                               52      });
    ▼ ▦ C:/EsriPress/GTKWebGIS/Chapter10    53
                                            54      view.then(function(view) {
       ▮ webmap_debug.html                  55          ▮console.▮log(view.map.layers.items[0].title);
  ▶ ⟲ js.arcgis.com                         56          console.log(view.map.layers.items[0].url);
                                            57          console.log(view.map.layers.items[0].layerId);
                                            58
```

A blue mark appears, which indicates there is a breakpoint at this line.

11. In your browser, reload the webmap_debug.html page. In the Sources window, you will see the code stops execution at the breakpoint.

12. In Developers Tools, click the Console tab, type **view.map.layers.items[0] .title**, press Enter, and note the result is Tectonic_Plates.

```
>  view.map.layers.items[0].title
<· "Tectonic_Plates"
```

This step illustrates that after a map view is loaded, the view's map property has information about the web map. The map's layers property has information about the operational layers of the web map. The first item in the layers is the Tectonic_Plates layer.

13. Still in the Console, type the following expressions, press Enter after each expression, and review the result of each action.

- **view.map.layers.items[0].url**

The result is the REST URL of the Tectonic_Plates feature service.

- **view.map.layers.items[0].layerId**

The result is 0, which is the order of the layer in the feature service.

- **view.map.layers.items[0].url + "/" + view.map.layers.items[0].layerId**

The result is the REST URL of the Tectonic_Plates feature layer.

- **view.map.layers.items[1].url + "/" + view.map.layers.items[1].layerId**

The result is the rest URL of the Earthquakes feature layer. You will use this URL in a query task in section 10.6.

```
> view.map.layers.items[0].title
< "Tectonic_Plates"
> view.map.layers.items[0].url
< "http://services2.arcgis.com/No7KRrFgpO516cMP/arcgis/rest/services/Tectonic_Plates/FeatureServer"
> view.map.layers.items[0].layerId
< 0
> view.map.layers.items[0].url + "/" + view.map.layers.items[0].layerId
< "http://services2.arcgis.com/No7KRrFgpO516cMP/arcgis/rest/services/Tectonic_Plates/FeatureServer/0"
> view.map.layers.items[1].url + "/" + view.map.layers.items[1].layerId
< "http://services2.arcgis.com/No7KRrFgpO516cMP/arcgis/rest/services/Earthquakes_2016_1_18/FeatureServer/0"
```

This step illustrates that when your code stops at a breakpoint, you can check the current values of your variables and expressions. This capability can help you understand the status of your code.

14. Click Sources, click line number 55 of webmap_debug.html. Note the blue mark disappears, which means the breakpoint has been removed.

15. With the source code of webmap_debug.html open, read the console.log lines.

Console.log can write expression values in the console without having to use breakpoints. Console.log is typically used in the development stage of your code and should be removed or commented out (turn it into a comment) before the code is deployed for production.

16.   Press the F8 key to continue the execution of your code.

17.   Click the Console tab and read the results of the console.log lines.

Next, you will review the HTTP traffic between your web browser and the GIS server, which is ArcGIS Online in this case.

18.   Click the Network tab, scroll up to see the map tile requests. Click a map tile request and prereview the response.

19.   Scroll to the bottom of the requests list, click the last query request, and preview the response.

This request retrieved all the features and attributes of the Earthquakes layer for browser-side display.

20. Right-click the last query request, and click Open in new tab.

The URL opens in a new tab, and the response displays in JSON format.

21. In the page URL, change f=json to **f=html**, press Enter, and note the ArcGIS REST Services Directory page displays.

This request illustrates how JavaScript API relies on ArcGIS REST API to communicate with GIS servers. While we used f=html to request the response in HTML format for us to understand, JavaScript API uses f=json to request the response in JSON format.

If you have a failed query, this page will help you determine what is wrong. You can experiment with the "where" clause, check the result, and find the correct SQL expression to use in your JavaScript.

In this section, you learned how to read messages in the console, set up breakpoints, step through JavaScript, examine expressions, and review HTTP traffic. These skills can help you quickly discover the cause of any issues in your code and provide you hints to solutions.

10.4 Monitor property changes

In this section, you will monitor the property changes as the 2D map view pans, zooms, and rotates, and make the 3D scene view follow the 2D view.

📖 **Note:** This section involves a lot of typing. You can avoid typing by copying code snippets that you will need from C:\EsriPress\GTKWebGIS\Chapter10\2d3d_Views_key.html. If you run into errors in this section, refer to the key file, or debug your code using the skills you just learned.

1. Start your web browser, and open C:\EsriPress\GTKWebGIS\ Chapter10\2d3d_Views.html. Notice the app displays a 2D view and a 3D view of the same layers.

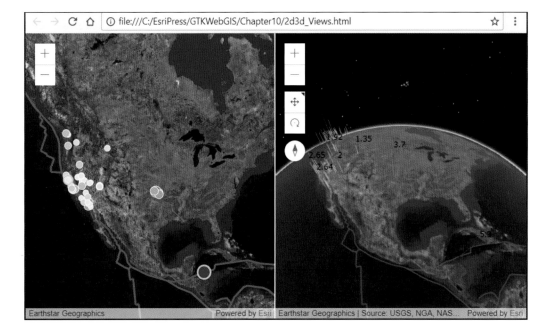

2. Pan, zoom, and rotate the 2D view, and notice that the 3D view doesn't follow.

3. Open C:\EsriPress\GTKWebGIS\Chapter10\2d3d_Views.html with Notepad++ or another JavaScript editor.

To link the 2D and 3D views, you must access the 3D view variable in the create_2dView function. Currently, the 3D view is a local variable and cannot be accessed from the create_2dView function. You will change the scope of the 3D view variable in the next step.

4. Add the following highlighted line above the `create_2dView()` line:

```
var view_2d, view_3d;
```

This line moves the view_2d and view_3d out of their local functions, so that these two variables can be accessed by other functions.

5. Change

```
var view = new MapView({
to
view_2d = new MapView({
```

By removing var at the beginning of the line and changing the variable name from view to view_2d, you are initializing the variable view_2d defined earlier instead of creating a local variable.

6. Similarly, change

```
var view = new SceneView({
to
view_3d = new SceneView({
```

7. Add the following highlighted lines to the `create_2dView` function:

```
    container: "viewDiv_2d"
});
view_2d.when(function() {
    view_2d.watch("extent", function(response){
        if (response){
            view_3d.center = response.center;
        }
    });
});
```

The code `view_2d.when` means that when the map of the 2D view has loaded, the event will trigger the function defined. The code `view_2d.watch` monitors the extent change of the 2D view. When the 2D view changes extent, the code triggers another function. The new extent is stored in the `response` argument. The code `view_3d.center = response.center;` uses the center of the 2D view to update the center of the 3D view.

8. Click the Save button in Notepad++.

9.   In your web browser, refresh the page C:\EsriPress\GTKWebGIS\
     Chapter10\2d3d_Views.html.

10. Pan the 2D map view and notice that the 3D scene view follows.

Next, you will program the 3D view to follow the scale change of the 2D view.

11. Add the following highlighted lines.

```
                    view_3d.center = response.center;
          }
    });
    view_2d.watch("scale", function(response){
          if (response){
                    view_3d.scale = response;
          }
    });
});
```

The new lines monitor the 2D view's scale changes. When the scale changes, the code sets the new scale of the 2D view to the scale of the 3D view.

12. Save the file in Notepad++.

13. In your web browser, refresh the page C:\EsriPress\GTKWebGIS\
     Chapter10\2d3d_views.html.

14. In the 2D view, zoom in and out, and observe that the 3D view follows.

Notice that 3D views do not have a consistent scale across the view. The actual scale and extent of the 3D view vary with the camera position and tilt angle. You can change the camera position and tilt angle of the 3D view manually (using right-click and drag).

Next, you will change the code so that when you rotate the 2D View, the 3D view will follow.

15. In Notepad++ in 2d3d_Views.html, add the following highlighted lines:

```
                    view_3d.scale = response;
          }
    });
    view_2d.watch("rotation", function(response){
          if (response){
                    view_3d.goTo({
```

```
                            heading: 0 - response
                  });
            }
      });
   });
```

The new lines of source code watch the change of the 2D view's rotation. The heading of the 3D view is a property of the view's viewing camera. The rotation of the 2D view and the camera heading of the 3D view measure in opposite directions. The code 0 — response corrects the direction differences so that the two views rotate in the same direction.

16.   Save the file in Notepad++.

17.   In your web browser, refresh the page C:\EsriPress\GTKWebGIS\Chapter10\ 2d3d_views.html.

18.   In the 2D view, right-click and drag to rotate the map, and notice that the 3D view follows.

In this section, you linked the 2D and 3D views by monitoring the 2D view's extent, scale, and rotation properties, and then synchronized them with the center, scale, and camera heading properties of the 3D view.

10.5 Incorporate widgets in your app

This section will add the legend widget to the app you created in the previous section.
   Note: This section involves a lot of typing. You can avoid typing by copying code snippets you need from C:\EsriPress\GTKWebGIS\Chapter10\2d3d_Views_legend_key.html. If you run into errors in this section, refer to the key file, or debug your code using the skills you just learned.

1.   Start Notepad++, open C:\EsriPress\GTKWebGIS\Chapter10\2d3d_Views_ legend.html.

This code is what you developed in the last section.

2.   Add the highlighted code to import the "esri/widgets/Legend" widget.

```
        "esri/WebScene",
        "esri/widgets/Legend",
    "dojo/domReady!"
  ],
  function(
  MapView, SceneView, WebMap, WebScene, Legend
  ) {
```

Notes:
- Make sure the order of Legend in the list of parameters corresponds to the order of "esri/ widgets/Legend" in the list of modules.
- Make sure you have the correct commas. In a JSON array, the elements are separated by commas, and the last element shouldn't have a following comma.

3.   Add the lines into the source code.

```
        view_2d.when(function() {
                var legend = new Legend({
                        id: "legend_2d",
                        view: view_2d
                })
                view_2d.ui.add(legend, "bottom-right");
```

The new code creates a legend object, associates it with view_2d, and positions the legend at the bottom right corner of the 2D view.

4.   Save your source code.

5.   In your web browser, reload C:\EsriPress\GTKWebGIS\Chapter10\2d3d_ Views_legend.html, and note a legend displays in the 2D view.

Next, you will add a legend to the 3D view.

For the Designer: for some reason, when the code is highlighted, our styles pane makes the size smaller, but that is not intentional. It should be the same size as everything else when it goes into design.

6. Add the highlighted code into the create_3dView function.

```
            container: "viewDiv_3d"
    });

    view_3d.when(function() {
        var legend3d = new Legend({
            id: "legend_3d",
            view: view_3d
        })
        view_3d.ui.add(legend3d, "bottom-right");
    });
```

The new source code monitors the 3D view. In the 3D view, the new code will execute a function, which creates a legend object, associates the legend with view_3d, and positions the legend at the bottom right corner of the 3D view.

7. Save your source code.

8. In your web browser, reload C:\EsriPress\GTKWebGIS\Chapter10\2d3d_Views_legend.html, and notice the legends in the 2D view and the 3D view.

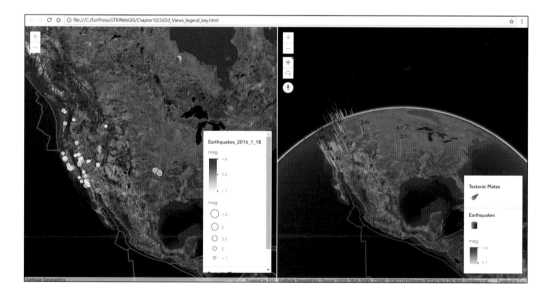

10.6 Work with tasks and promises

QueryTask is a commonly used task to perform spatial and attribute queries on a layer. In this section, you will create an app that allows users to define some query parameters, execute the query, and display the results in the 2D and the 3D views.

📋 **Note:** This section involves a lot of typing. You can avoid typing by copying code snippets you need from C:\EsriPress\GTKWebGIS\Chapter10\2d3d_query_key.html. If you run into errors in this section, refer to the key file, or debug your code using the skills you just learned.

1. Start Notepad++, and open C:\EsriPress\GTKWebGIS\Chapter10\2d3d_ query.html.

This is the code you developed in the last section.

2. In the HTML body, add the highlighted source code.

```
<body>
    <div id="viewDiv_2d"></div>
    <div id="viewDiv_3d"></div>
    <div id="optionsDiv">
        Magnitude
        <select id="signSelect">
            <option value=">">is greater than</option>
            <option value="<">is less than</option>
            <option value="=">is equal to</option>
        </select>
        <input id="valSelect" value="3" />
        <br>
        <br>
        <button id="doBtn">Execute</button>
        <br>
        <p><span id="printResults"></span></p>
    </div>
</body>
```

The new code adds `optionsDiv` div, an HTML container that includes the following HTML elements:

- A `Magnitude` label.
- A drop-down menu of comparison operators.
- An input box for users to type the earthquake magnitude to search. The default value is 3.

- A button for users to click to execute the query task.
- An HTML span to display information about the query result.

3. In the style section of the source code, add the following highlighted code:

```
#optionsDiv {
 position: absolute;
 top: 0px;
 right: 0px;
 max-width: 450px;
 background-color: dimgray;
 color: white;
 text-align: center;
 z-index: 30;
 padding: 10px;
 border-radius: 5px;
 }
</style>
```

The new CSS code positions the optionsDiv division in the upper-right corner of the page, and defines its max-width, color, alignment, z-index (so that the div displays on top of the 3D view), padding, and border-radius.

4. Save your code.

5. Open C:\EsriPress\GTKWebGIS\Chapter10\2d3d_query.html in the browser and review the new HTML elements added.

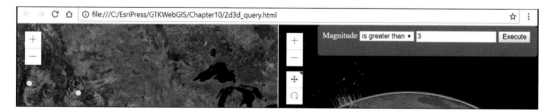

6. In Notepad++, add the following highlighted code:

```
require([
"esri/views/MapView",
        "esri/views/SceneView",
        "esri/WebMap",
        "esri/WebScene",
```

```
        "esri/widgets/Legend",
        "esri/layers/GraphicsLayer",
        "esri/symbols/SimpleMarkerSymbol",
        "esri/symbols/PointSymbol3D",
        "esri/symbols/ObjectSymbol3DLayer",
        "esri/tasks/QueryTask",
        "esri/tasks/support/Query",
        "dojo/dom",
        "dojo/on",
        "dojo/_base/array",
    "dojo/domReady!"
  ],
  function(
  MapView, SceneView, WebMap, WebScene, Legend, GraphicsLayer,
SimpleMarkerSymbol, PointSymbol3D, ObjectSymbol3DLayer, QueryTask,
Query, dom, on, arrayUtils
      ) {
```

Typically, developers add the modules needed step by step. To simplify the tutorial, we added the necessary modules all at once. You must ensure that the order of the modules corresponds to the order of the parameters, and ensure you have the correct commas.

- "esri/layers/GraphicsLayer" will be used to display the result of the earthquake query.
- "esri/symbols/SimpleMarkerSymbol" will be used to style the matching earthquakes in the 2D view.
- "esri/symbols/PointSymbol3D" and "esri/symbols/ObjectSymbol3DLayer" will be used to style the matching earthquakes in the 3D view.
- "esri/tasks/QueryTask" and "esri/tasks/support/Query" will be used to create the query task and query parameters.
- "dojo/dom" will be used to find the execute button.
- "dojo/on" will be used to define an event handler for the execute button.
- "dojo/_base/array" will be used to process the array of earthquakes in the query result.

7.   In Notepad++, add the following highlighted code:

```
var view_2d, view_3d;
var results2DLyr = new GraphicsLayer();
var results3DLyr = new GraphicsLayer();
```

The new code creates two graphics layers for displaying the query result in the 2D view and 3D view respectively.

Next, you will add the two layers into the 2D view and the 3D view.

8. In Notepad++, add the following highlighted code:

```
view_2d.when(function() {
        webmap.add(results2DLyr);
```

9. In Notepad++, add the following highlighted code:

```
view_3d.when(function() {
        scene.add(results3DLyr);
```

10. Add the following highlighted source code:

```
    create_3dView();
on(dom.byId("doBtn"),"click", doQuery);
```

The new code creates an event listener. When users click the execute button, the code will execute the doQuery function.

11. Add the following highlighted source code to create the doQuery function:

```
            view_3d.ui.add(legend3d, "bottom-right");
    });
    }

    function doQuery(){
            var featureLayerUrl = view_2d.map.layers.items[1].
url + "/" + view_2d.map.layers.items[1].layerId;
            var qTask = new QueryTask({
                url: featureLayerUrl
            });
            var params = new Query({
                returnGeometry: true,
                outFields: ["*"]
            });
            var expressionSign = dom.byId("signSelect");
            var value = dom.byId("valSelect");
            params.where = "mag" + expressionSign.value +
value.value;

            qTask.execute(params)
                    .then(getResults)
                    .otherwise(promiseRejected);
        }
```

10

```
  });
</script>
```

The new code has the following logic:

- Find the URL of the Earthquakes feature layer in the web map. Refer to section 10.3 to understand that the URL is built.
- Create a query task pointing to this URL.
- Define query parameters to return the geometries and all attribute fields of the matching earthquakes.
- Build a where clause by concatenating the mag field, the comparison operator users select, and the magnitude users specify.
- Add the where clause as the where parameter for the query.
- Execute the query task using the parameters defined earlier, call getResults() once the query succeeds, and call promiseRejected() if the query fails.

12. Add the following highlighted code to create the getResults function and the promiseRejected function:

```
    .otherwise(promiseRejected);
  }

  function getResults(response) {
        // print the number of results returned to the user
        dom.byId("printResults").innerHTML = response.features.
length + " result(s) found!";
  }

  function promiseRejected(err) {
        console.error("Query failed: ", err.message);
  }
```

The getResults function will display the number of matching earthquakes in the printResults span that was added in step 2 of this section. It there are any error messages, the promiseRejected function will display the error messages in the browser console.

13. Save your code.

14. In your web browser, reload C:\EsriPress\GTKWebGIS\Chapter10\2d3d_query.html.

15. Click the Execute button and note the number of matching results displayed.

If you don't see the number of results, you can check your web browser console for errors, and refer to 2d3d_query_key.html to fix your code.

Next, you will add code to display the results in the 2D view.

16. Add the following highlighted code to create the displayResultsIn2D function.

```
        console.error("Query failed: ", err.message);
  }
  function displayResultsIn2D(response) {
        results2DLyr.removeAll();
        var featureResults2D = arrayUtils.map(response.features,
function(feature) {
                feature.symbol = new SimpleMarkerSymbol({
                        style: "circle",
                        color: "yellow",
                        size: "8px",
                        outline: {
                                color: [ 255, 255, 0 ],
                                width: 6
                        }
                });
                return feature;
        });

        results2DLyr.addMany(featureResults2D);
        // animate to the results after they are added to the map
        view_2d.goTo(featureResults2D);
  }
```

The displayResultsIn2D function has the following logic:
- Clear the graphics of the previous results from the results2DLyr layer.
- Loop through the earthquakes in the results. For each earthquake, create a point marker symbol with a green circle, and assign the symbol to the earthquake.
- Add the earthquakes (along with their symbols) to the results2DLyr graphics layer.
- Zoom the map to the extent that includes all the earthquakes in the result.

17. Add the following highlighted code to call the displayResultsIn2D function:

```
function getResults(response) {
        dom.byId("printResults").innerHTML = response.features.
length + " result(s) found!";
        displayResultsIn2D(response);
}
```

18. Save your code.

19. In your browser, reload C:\EsriPress\GTKWebGIS\Chapter10\2d3d_query.html.

20. Click the Execute button and note how the matching earthquakes are displayed in the 2D view.

Next, you will add code to display the results in the 3D view.

21. Add the following highlighted code to create the displayResultsIn3D function:

```
        view_2d.goTo(featureResults2D);
}
function displayResultsIn3D(response) {
        results3DLyr.removeAll();
        var featureResults3D = arrayUtils.map(response.features,
function(feature) {
                var newFeature = feature.clone();
                        newFeature.symbol = new PointSymbol3D({
                                symbolLayers: [new
ObjectSymbol3DLayer({
                                        material: {
                                                color: "green"
                                        },
                                        resource: {
                                                primitive: "cone"
                                        },
                                        width: 300000,
                                        height: 1000000
                                })]
                        });
                return newFeature;
        });
```

```
results3DLyr.addMany(featureResults3D);
}
```

The `displayResultsIn3D` function has the following logic:
- Clear the graphics of the previous results from the results3DLyr layer.
- Loop through the earthquakes in the results.
  - For each earthquake, clone the earthquake, so the 3D symbol below will be given to the earthquake clone and not the earthquake that is already displayed in the 2D view.
  - For each earthquake, create a 3D point symbol with a green cone and assign the symbol to the earthquake clone.
- Add the earthquake clones (along with their symbols) to the results3DLyr graphics layer.

22. Add the following highlighted code to call the displayResultsIn3D function:

```
displayResultsIn2D(response);
displayResultsIn3D(response);
}
```

23. Save your code.

24. In your browser, reload C:\EsriPress\GTKWebGIS\Chapter10\2d3d_query. html.

25. Change the default magnitude value to **5.5**, click the Execute button, and note the matching earthquakes are displayed in the 2D and the 3D views.

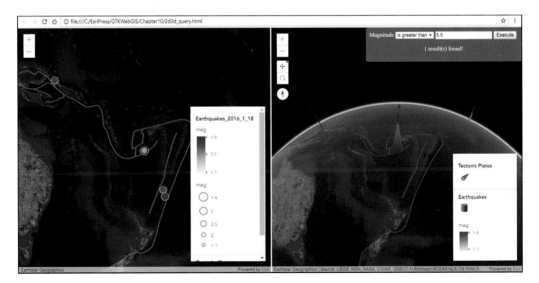

In this tutorial, you learned the general steps to create a JavaScript app using JavaScript API. You explored loading layers via web maps and web scenes. You then learned the basic skills to debug JavaScript code. You further learned how to handle events, integrate widgets, and perform queries. The apps created in this tutorial are simple, but they have demonstrated the flexibility, potential, and fun of JavaScript API.

Programming can be challenging. Sometimes a tiny typo can take hours to discover and fix. If you felt frustrated while completing this tutorial, you are not alone. The initial frustration is normal for everyone who is learning something new. It's a sign of true learning. After you get through the initial frustration, you can start building interesting and amazing apps. You can further your journey to Web GIS programming by exploring more JavaScript samples, using what you learned in your projects, turning your creativity into real software products, and advancing the application and development of Web GIS.

QUESTIONS AND ANSWERS

1. I changed my JavaScript code and refreshed the page in my web browser. However, I cannot see the latest changes in the browser. What is wrong? How do I fix it?

   Answer: Browser caching is causing the problem. Browsers often cache a version of the web pages you visited and use this cached version to speed up performance. However, this cache can cause browsers to use an outdated version of an updated page. If you have this problem, search online for proper browser settings.

2. In my web browser, I use the following special URL: file:///C:/EsriPress/ GTKWebGIS/Chapter10/2d3d_Views_key.html. Is this okay?

   Answer: Yes, you can view and develop simple JavaScript apps using the file URL. You should deploy official web apps and complex web apps to a web server and access these apps via HTTP or HTTPS URLs.

3. **I shared my app URL (http://localhost/WebGIS/app.html) with my users. However, my users said they could not see my app. Why?**

   Answer: Localhost is relative and points to the computer that the users are currently using. Therefore, whenever you go to this URL on your computer, this URL points to your computer. If users go to this URL on their computers, then the URL points to their computers, which do not have your app deployed. So, users will not have access to your app.

   Although localhost is convenient for running your app on your own computer, when you share your app URL with your audience, use the correct host name instead of localhost.

4. **Microsoft IIS is a commonly used web server. How do I know whether I have IIS installed on my Windows computer? How do I install IIS?**

   Answer: To verify whether your Windows computer has IIS installed, simply open a web browser and type **http://localhost** or **http://your_computer_name**. Typically, a web page appears, which means that IIS has been installed. Otherwise, search online for help documents that explain how to install IIS.

5. **How do I deploy a JavaScript web app to IIS?**

   Answer: You simply copy your web app files to a folder under C:\inetpub\ wwwroot and give everyone read access to the folder. Search for online help documents that explain the details.

ASSIGNMENT

Assignment 10: Develop a web app using JavaScript API.

Requirements:
- Find a sample or multiple samples on the JavaScript API website. Adapt the sample(s).
- Use a web map/scene or layers different from the original sample(s). You can use the web maps, scenes, or layers you created before, or search on ArcGIS Online or ArcGIS Open Data (**http://opendata.arcgis.com**).

- Incorporate a query task, geoprocessor, or other type of task.
- Incorporate a widget.

What to submit:
- The URL of the original sample(s) with which you started
- Your new source code

Resources

"2D Visualization with the ArcGIS API for JavaScript," Kristian Ekenes and Jeremy Bartley, https://www
.youtube.com/watch?v=5K-O6BaKNwo (or http://bit.ly/2iLoSMg).

"ArcGIS API for JavaScript," https://developers.arcgis.com/javascript.

"ArcGIS API for JavaScript: API reference," https://developers.arcgis.com/javascript/latest/api-reference/
index.html (or http://arcg.is/2mY1FLl).

"ArcGIS API for JavaScript: Getting Started," Bjorn Svensson and Undral Batsukh, https://www.youtube
.com/watch?v=pYHWoSNsSIU (or http://bit.ly/2A0yn3P).

"ArcGIS API for JavaScript: sample code," https://developers.arcgis.com/javascript/latest/sample-code/
index.html (or http://arcg.is/2i11o8M).

"ArcGIS API for JavaScript: Tips and Tricks for Developing and Debugging Apps," Heather Gonzago and
Kelly Hutchins, https://www.youtube.com/watch?v=nVMeu65qnc4 (http://bit.ly/2A4fddm).

"ArcGIS DevLabs," https://developers.arcgis.com/labs/develop/index.html#javascript.

"Basics of JavaScript Web Apps," https://www.esri.com/training/catalog/580fc1dea4a46d172b116049/
basics-of-javascript-web-apps (or http://arcg.is/2nXm5mh).

"Developing Web Apps with ArcGIS API for JavaScript," https://www.esri.com/training/catalog/
57630437851d31e02a43f267/developing-web-apps-with-arcgis-api-for-javascript (or http://arcg.is/
2mr5OCx).

"GeoNet Forum for ArcGIS API for JavaScript," https://community.esri.com/community/developers/
web-developers/arcgis-api-for-javascript (or http://arcg.is/2AsAEWk).

"Working in 3D with the ArcGIS API for JavaScript," Jesse van den Kieboom and Adrian Blumer, https://
www.youtube.com/watch?v=7aFPVvS9-dk (or http://bit.ly/2mYydEP).

"Working with promises," https://developers.arcgis.com/javascript/latest/guide/working-with-promises/
index.html (or http://arcg.is/2BiadzB).

"Working with properties," https://developers.arcgis.com/javascript/latest/guide/working-with-props/
index.html (or http://arcg.is/2A0W0tc).

APPENDIX A

Image credits

Chapter 1

Esri; Esri; Esri; Esri; Esri; Esri; Esri; Esri; Left side: DC GIS, DDOT, DRES, GSA, OCTO, NHD, VITA, METI/NASA, Esri, DeLorme, HERE, USGS, USDA, EPA, EarthSat, Intermap, IPC, and Tom-Tom, Photos by Allen Carroll, Esri. Right side: City of Redlands, City of Riverside, County of Riverside, Esri, HERE, DeLorme, Intermap, iPC, USGS, USDA, and EPA. Photo by Pinde Fu; Esri; Esri; Esri; Esri; Esri; Esri; Esri; Esri; Esri; Esri; City of Redlands, County of Riverside, Bureau of Land Management, Esri, HERE, Garmin, Increment P, Intermap, USGS, METI/NASA, NGA, EPA, USDA; Esri; Esri; Esri; Esri; Esri; Esri; Esri; Esri; Esri; Esri.

Chapter 2

Esri; Esri; Esri; Esri; Esri; Esri and USGS; Esri; Esri; Esri; Esri; Census; Esri; Esri; Esri; Esri, HERE, Garmin, FAO, USGS, EPA, NPS, Census; Esri, HERE, Garmin, FAO, USGS, EPA, NPS, Census; Esri; Esri; Esri; Esri; Esri, HERE, Garmin, NGA, USGS, Census; Esri; Esri; Esri; Esri; Esri; Esri; Esri; Esri; Esri; Esri; Census; Esri; Esri; Census; Esri; Esri; Esri; Esri; Esri; Esri, Census; Esri; Esri; Esri, HERE, Garmin, NGA, USGS | Esri, US Census Bureau, Infogroup | Esri, HERE; Esri, HERE, Garmin, NGA, USGS, US Census Bureau, Infogroup; County and City of Denver, Esri, HERE, Garmin, NGA, USGS, US Census Bureau, Infogroup; Esri; Esri; Esri

Chapter 3

Esri; USGS, National Atlas, NOAA National Climatic Data Center, Esri, FAO, NOAA, NRCan; Esri; Esri; Esri; Esri; Esri; USGS, National Atlas, NOAA National Climatic Data Center; Esri; Esri; Esri; Esri; Esri; Esri; Esri; Esri; Esri; Esri; Esri, FAO, NOAA, NRCan; USGS, National Atlas, NOAA

National Climatic Data Center, Esri, FAO, NOAA, NRCan; Esri; Esri; Esri; Esri; Esri; Esri; Esri; Esri; Esri; Esri; Esri; Esri; Esri; Esri; Esri; Esri; Esri; Esri; USGS, Esri, FAO, NOAA, NRCan; Esri; Esri; Esri; Esri; Esri; USGS, National Atlas, NOAA National Climatic Data Center, Esri, FAO, NOAA, NRCan; Esri; Esri; Esri; Esri; Esri, HERE, Garmin, NGA, USGS, National Atlas, NOAA National Climatic Data Center

Chapter 4

Esri; Esri; Esri; Esri; Esri; Esri; Esri; Esri; Esri; Esri; Esri; Esri; Esri; Esri; Bureau of Land Management, Esri, HERE, Garmin, NGA, USGS, NPS; Esri; Esri; Esri; Esri; Esri; Esri; Esri; Esri; Esri; City of Redlands, County of Riverside, Bureau of Land Management, Esri, HERE, Garmin, Increment P, Intermap, USGS, EPA, USDA; Esri; City of Redlands, County of Riverside, Bureau of Land Management, Esri, HERE, Garmin, Increment P, Intermap, USGS, EPA, USDA; Esri; Esri; Esri; Esri; Esri; Esri; Esri; Esri; Esri; Esri; Esri; City of Redlands, County of Riverside, Bureau of Land Management, Esri, HERE, Garmin, Increment P, Intermap, USGS, EPA, USDA; Esri; Esri; Esri; The Android robot is reproduced or modified from work created and shared by Google and used according to terms described in the Creative Commons 3.0 Attribution License; Esri; Esri

Chapter 5

Esri; Esri; Esri; Esri; Esri; Esri; Esri; Esri, USGS, FAO, NOAA, National Atlas, NOAA National Climatic Data Center; Esri. Data from USGS, National Atlas, NOAA National Climatic Data Center

Chapter 6

Esri; Esri; Esri; US Department of Transportation; Esri; Esri; Esri; Esri; Esri; Esri; Esri, HERE, Garmin, NGA, USGS, NPS | NOAA/NOS/OCS nowCOAST, NOAA/NWS and NOAA/OAR/NSSL | NOAA/NWS | NOAA, Esri | Esri, HERE, Garmin; Esri, USGS, NOAA/NWS, DeLorme, USGS, NPS; DCGIS, VITA, Esri, HERE, Garmin, FAO, NOAA, USGS, EPA, NPS, Esri; Esri; Esri; Esri; Esri; City of Redlands, County of Riverside, San Bernardino County, Bureau of Land Management, Esri, HERE, Garmin, Increment P, Intermap, USGS, METI/NASA, EPA, USDA, California Department of Transportation; Esri; Esri; Esri; Esri; Esri; Esri; City of Redlands, County of Riverside, San Bernardino County, Bureau of Land Management, Esri, HERE, Garmin, Increment P, Intermap, USGS, METI/NASA, EPA, USDA, California Department of Transportation; Esri; Esri; Esri; Esri; Esri; Esri; Esri; Esri; Esri; Esri; Esri; Esri; Esri; City of Redlands, County of Riverside, San Bernardino County, Bureau of Land Management, Esri, HERE, Garmin, Increment P, Intermap, USGS, METI/NASA, NGA, EPA, USDA; Esri; Esri; Esri; City of Redlands, County of Riverside, San Bernardino County, Bureau of Land Management, Esri, HERE, Garmin, Increment P, Intermap, USGS, METI/NASA, NGA, EPA, USDA; US Census; US Census; Esri; Esri; Esri; Esri; Esri; Esri; US Census, Esri, USGS | Esri, HERE, Garmin, FAO, NOAA, USGS, EPA;

Chapter 7

Esri; Left: USDA FSA, DigitalGlobe, GeoEye, Microsoft, City of Montreal, Canada, Esri Canada, Esri. Right: Esri, DeLorme, FAO, NOAA, USGS, EPA, and US Census; Esri; Esri; NOAA; City of Venice City Planning Data Portal, Esri, HERE, Garmin, Increment P, USGS | USGS, NGA, NASA, CGIAR, GEBCO, N Robinson, NCEAS, NLS, OS, NMA, Geodatastyrelsen and the GIS User Community; Pix4D, Ascending Technologies, and Esri; Esri; Pictometry International, Microsoft | USGS, NGA, NASA, CGIAR, GEBCO, N Robinson, NCEAS, NLS, OS, NMA, Geodatastyrelsen and the GIS User Community; USDA FSA, Digital Global, GeoEye, Microsoft, CNES/Airbus DS; Esri; Esri; Esri; Esri; Esri; Microsoft | USGS, NGA, NASA, CGIAR, GEBCO, N Robinson, NCEAS, NLS, OS, NMA, Geodatastyrelsen and the GIS User Community; Esri; Esri; Esri; Henan University; Esri; Esri; Esri; Esri; Esri; Esri; Esri; Esri; City of Rancho Cucamonga, Microsoft | USGS, NGA, NASA, CGIAR, GEBCO, N Robinson, NCEAS, NLS, OS, NMA, Geodatastyrelsen and the GIS User Community | City of Rancho Cucamonga, San Bernardino County, Esri, HERE, Garmin, Increment P, USGS, NPS, EPA, US Census Bureau, USDA, Bureau of Land Management; Esri; Esri; Esri; City of Rancho Cucamonga, Microsoft | USGS, NGA, NASA, CGIAR, GEBCO, N Robinson, NCEAS, NLS, OS, NMA, Geodatastyrelsen and the GIS User Community | City of Rancho Cucamonga, San Bernardino County, Esri, HERE, Garmin, Increment P, USGS, NPS, EPA, US Census Bureau, USDA, Bureau of Land Management; US Bureau Transportation Statistics, ArcWorld; Esri; Esri; Map data © OpenStreetMap contributors, CC-BY-SA | USGS, NGA, NASA, CGIAR, GEBCO, N Robinson, NCEAS, NLS, OS, NMA, Geodatastyrelsen and the GIS User Community; Esri; Esri; City of Rancho Cucamonga, Microsoft | USGS, NGA, NASA, CGIAR, GEBCO, N Robinson, NCEAS, NLS, OS, NMA, Geodatastyrelsen and the GIS User Community | City of Rancho Cucamonga, San Bernardino County, Esri, HERE, Garmin, Increment P, USGS, NPS, EPA, US Census Bureau, USDA, Bureau of Land Management; City of Rancho Cucamonga, Microsoft | USGS, NGA, NASA, CGIAR, GEBCO, N Robinson, NCEAS, NLS, OS, NMA, Geodatastyrelsen and the GIS User Community | City of Rancho Cucamonga, San Bernardino County, Esri, HERE, Garmin, Increment P, USGS, NPS, EPA, US Census Bureau, USDA, Bureau of Land Management; Esri; City of Rancho Cucamonga, Microsoft | USGS, NGA, NASA, CGIAR, GEBCO, N Robinson, NCEAS, NLS, OS, NMA, Geodatastyrelsen and the GIS User Community | City of Rancho Cucamonga, San Bernardino County, Esri, HERE, Garmin, Increment P, USGS, NPS, EPA, US Census Bureau, USDA, Bureau of Land Management; Esri; Esri; Esri; City of Venice City Planning Data Portal, Esri, HERE, Garmin, Increment P, USGS | USGS, NGA, NASA, CGIAR, GEBCO, N Robinson, NCEAS, NLS, OS, NMA, Geodatastyrelsen and the GIS User Community; Esri

Chapter 8

Esri; Esri; Esri; Esri; Esri; Esri; NYC Taxi & Limousine Commission, City of New York; Esri; Esri; Esri; Esri; City and County of San Francisco, Esri; Georgia Power; Esri; Esri; Esri; Esri; County of Los Angeles, Esri, HERE, Garmin, FAO, USGS, Bureau of Land Management, EPA, NPS; Esri; Esri; Esri; County of Riverside, Esri, HERE, Garmin, USGS, Bureau of Land Management, EPA, NPS, USDA; Esri; Esri; Esri; Esri; Esri, HERE, Garmin, USGS, Bureau of Land Management, EPA, NPS, USDA; Esri; Esri; Esri; Esri; Esri; Esri; Esri; Esri; Esri; Esri; Esri; Courtesy of US Bureau

Transportation Statistics, ArcWorld; Esri; Esri; Esri; Esri; Esri; Esri; Esri; Esri; Esri; US Bureau Transportation Statistics, ArcWorld, Esri, HERE, Garmin, FAO, USGS, EPA, NPS; US Bureau Transportation Statistics, ArcWorld, Esri, HERE, Garmin, FAO, USGS, EPA, NPS; Esri; Esri; Esri; Esri; US Bureau Transportation Statistics, ArcWorld, Esri, HERE, Garmin, FAO, USGS, EPA, NPS | Esri, HERE, Garmin, FAO, USGS, EPA, NPS; US Bureau Transportation Statistics, ArcWorld, Esri, HERE, Garmin, FAO, USGS, EPA, NPS | Esri, HERE, Garmin, FAO, USGS, EPA, NPS; Esri; Esri; Esri; Esri; Esri; Esri; Esri

Chapter 9

Esri; Esri; Esri; Esri; Esri, USDA Farm Service Agency; Esri, USGS, AWS, NASA | Texas Parks & Wildlife, Esri, HERE, Garmin, USGS, EPA, NGA, EPA, USDA, NPS; Esri, GEBCO, DeLorme | Esri, GEBCO, IHO-IOC, GEBCO, NGS | HYCOM, Esri; Esri, USGS | NASA | Nebraska Game & Parks Commission, Esri, HERE, Garmin, FAO, NOAA, EPA; Esri; Esri, USDA Farm Service Agency; Esri, FEMA; MapmyIndia, Esri, HERE, DeLome, USGS, METI/NASA, NGA; Esri, HERE, Garmin, FAO, NOAA, USGS, EPA | Esri, USDA Farm Service Agency; Esri; Esri; Esri; City of Redlands, County of Riverside, San Bernardino County, Bureau of Land Management, Esri, HERE, Garmin, Increment P, Intermap, USGS, METI/NASA, EPA, USDA | Esri, USDA Farm Service Agency; City of Redlands, County of Riverside, San Bernardino County, Bureau of Land Management, Esri, HERE, Garmin, Increment P, Intermap, USGS, METI/NASA, EPA, USDA | Esri, USDA Farm Service Agency; Esri; Esri; City of Redlands, County of Riverside, San Bernardino County, Bureau of Land Management, Esri, HERE, Garmin, Increment P, Intermap, USGS, METI/NASA, EPA, USDA | Esri, USDA Farm Service Agency; City of Redlands, County of Riverside, San Bernardino County, Bureau of Land Management, Esri, HERE, Garmin, Increment P, Intermap, USGS, METI/NASA, EPA, USDA | Esri, USDA Farm Service Agency; City of Redlands, County of Riverside, San Bernardino County, Bureau of Land Management, Esri, HERE, Garmin, Increment P, Intermap, USGS, METI/NASA, EPA, USDA | Esri, USDA Farm Service Agency; City of Redlands, County of Riverside, San Bernardino County, Bureau of Land Management, Esri, HERE, Garmin, Increment P, Intermap, USGS, METI/NASA, EPA, USDA | Esri, USDA Farm Service Agency; City of Redlands, County of Riverside, San Bernardino County, Bureau of Land Management, Esri, HERE, Garmin, Increment P, Intermap, USGS, METI/NASA, EPA, USDA | Esri, USDA Farm Service Agency; City of Redlands, County of Riverside, San Bernardino County, Bureau of Land Management, Esri, HERE, Garmin, Increment P, Intermap, USGS, METI/NASA, EPA, USDA | Esri, USDA Farm Service Agency; City of Redlands, County of Riverside, San Bernardino County, Bureau of Land Management, Esri, HERE, Garmin, Increment P, Intermap, USGS, METI/NASA, EPA, USDA | Esri, USDA Farm Service Agency; City of Redlands, County of Riverside, San Bernardino County, Bureau of Land Management, Esri, HERE, Garmin, Increment P, Intermap, USGS, METI/NASA, EPA, USDA | Esri, USDA Farm Service Agency; City of Redlands, County of Riverside, San Bernardino County, Bureau of Land Management, Esri, HERE, Garmin, Increment P, Intermap, USGS, METI/NASA, EPA, USDA | Esri, USDA Farm Service Agency; City of Redlands, County of Riverside, San Bernardino County, Bureau of Land Management, Esri, HERE, Garmin, Increment P, Intermap, USGS, METI/NASA, EPA, USDA | Esri, USDA Farm Service Agency; Esri; Esri, HERE, DeLorme, Intermap, Increment P, GEBCO, USGS, FAO, NPS, NRCAN, GeoBase, IGN, Kadaster NL, Ordnance Survey, Esri Japan, METI, Esri China (Hong Kong), swisstopo, MapmyIndia, OpenStreetMap contributors, and the GIS User

Community; Esri; Esri; Esri, HERE, DeLorme, Intermap, Increment P, GEBCO, USGS, FAO, NPS, NRCAN, GeoBase, IGN, Kadaster NL, Ordnance Survey, Esri Japan, METI, Esri China (Hong Kong), swisstopo, MapmyIndia, OpenStreetMap contributors, and the GIS User Community; Esri; Esri; Esri; Esri; Esri; Esri; Esri, HERE, DeLorme, Intermap, Increment P, GEBCO, USGS, FAO, NPS, NRCAN, GeoBase, IGN, Kadaster NL, Ordnance Survey, Esri Japan, METI, Esri China (Hong Kong), swisstopo, MapmyIndia, OpenStreetMap contributors, and the GIS User Community; Esri; Esri; Esri; Esri; City of Redlands, County of Riverside, San Bernardino County, Bureau of Land Management, Esri, HERE, Garmin, Increment P, Intermap, USGS, METI/NASA, EPA, USDA; City of Redlands, County of Riverside, San Bernardino County, Bureau of Land Management, Esri, HERE, Garmin, Increment P, Intermap, USGS, METI/NASA, EPA, USDA; City of Redlands, County of Riverside, San Bernardino County, Bureau of Land Management, Esri, HERE, Garmin, Increment P, Intermap, USGS, METI/NASA, EPA, USDA; Esri; City of Redlands, County of Riverside, San Bernardino County, Bureau of Land Management, Esri, HERE, Garmin, Increment P, Intermap, USGS, METI/NASA, EPA, USDA; Esri; City of Redlands, County of Riverside, San Bernardino County, Bureau of Land Management, Esri, HERE, Garmin, Increment P, Intermap, USGS, METI/NASA, EPA, USDA | Esri, USDA Farm Service Agency; USGS

Chapter 10

Esri; Esri; Esri; Esri; Esri; Esri; Esri; Esri, HERE, Garmin, NGA, USGS | USGS, NGA, NASA, CGIAR, GEBCO, N Robinson, NCEAS, NLS, OS, NMA, Geodatastyrelsen and the GIS User Community; Esri; Esri; Esri; Esri; Esri; Esri; Esri; Esri; Esri; Esri; Esri, HERE, Garmin, NGA, USGS | USGS, NGA, NASA, CGIAR, GEBCO, N Robinson, NCEAS, NLS, OS, NMA, Geodatastyrelsen and the GIS User Community; Esri; Esri; Esri; Earthstar Geographics, USGS, NGA, NASA, Esri, CGIAR, GEBCO, N Robinson, NCEAS, NLS, OS, NMA, Geodatastyrelsen and the GIS User Community; Earthstar Geographics, USGS, NGA, NASA, Esri, CGIAR, GEBCO, N Robinson, NCEAS, NLS, OS, NMA, Geodatastyrelsen and the GIS User Community; Earthstar Geographics, USGS, NGA, NASA, Esri, CGIAR, GEBCO, N Robinson, NCEAS, NLS, OS, NMA, Geodatastyrelsen and the GIS User Community; Esri; Earthstar Geographics, USGS, NGA, NASA, Esri, CGIAR, GEBCO, N Robinson, NCEAS, NLS, OS, NMA, Geodatastyrelsen and the GIS User Community

APPENDIX B

Data credits

\\EsriPress\GTKWebGIS\Chapter1\Locations.CSV, courtesy of Esri.

\\EsriPress\GTKWebGIS\Chapter1\map_tour_thumbnail.jpg, courtesy of Esri.

\\EsriPress\GTKWebGIS\Chapter2\Top_50_US_Cities.CSV, courtesy of US Census Bureau.

\\EsriPress\GTKWebGIS\Chapter2\images\thumbnail.jpg, courtesy of Esri, Delorme, HERE, US Census Bureau.

\\EsriPress\GTKWebGIS\Chapter2\blue_down.png, courtesy of Esri.

\\EsriPress\GTKWebGIS\Chapter2\reasons-why-people-move.jpg, courtesy of US Census Bureau.

\\EsriPress\GTKWebGIS\Chapter3\USGS_earthquake_cause.gif, courtesy of USGS.

\\EsriPress\GTKWebGIS\Chapter4\311_Incidents.csv, courtesy of Esri.

\\EsriPress\GTKWebGIS\Chapter4\Survey_Icon.png, courtesy of Esri.

\\EsriPress\GTKWebGIS\Chapter5\Data.gdb\Earthquakes, courtesy of USGS.

\\EsriPress\GTKWebGIS\Chapter5\Data.gdb\Hurricanes, courtesy of NOAA National Climatic Data Center.

\\EsriPress\GTKWebGIS\Chapter5\Assignment.gdb\Wildfires2015_16, courtesy of USGS.

\\EsriPress\GTKWebGIS\Chapter5\Assignment.gdb\Tornadoes2015_16, courtesy of NOAA National Weather Service.

\\EsriPress\GTKWebGIS\Chapter5\US_Population_200_Years.csv, courtesy of US Census Bureau.

\\EsriPress\GTKWebGIS\Data\Chapter7\FunParkData\Park.gdb, courtesy of Esri.

\\EsriPress\GTKWebGIS\Data\Chapter7\FunParkData\rpk\Castle.rpk, courtesy of Esri.

\\EsriPress\GTKWebGIS\Data\Chapter7\FunParkData\rpk\Soccerfield.rpk, courtesy of Esri.

\\EsriPress\GTKWebGIS\Data\Chapter7\FunParkData\rpk\VeniceFacades.rpk, courtesy of Esri.

\\EsriPress\GTKWebGIS\Data\Chapter7\FunParkData\rpk\FunPark_Points_URL.txt, courtesy of Esri.

\\EsriPress\GTKWebGIS\Data\Chapter8\Lab_Data.gdb\point_lyr, courtesy of Esri.

\\EsriPress\GTKWebGIS\Data\Chapter8\Lab_Data.gdb\Railroads, courtesy of US Bureau of
    Transportation Statistics.

\\EsriPress\GTKWebGIS\Data\Chapter8\Lab_Data.gdb\Rivers, courtesy of ArcWorld.

\\EsriPress\GTKWebGIS\Chapter8\Assignment_Data\ToolData\rail100k.sdc.xml, courtesy of US Bureau
    of Transportation Statistics.

\\EsriPress\GTKWebGIS\Chapter8\Assignment_Data\ToolData\rivers.sdc.xml, courtesy of ArcWorld.

\\EsriPress\GTKWebGIS\Data\Chapter8\Planning.tbx, courtesy of Esri.

\\EsriPress\GTKWebGIS\Data\Chapter8\Planning.tbx\Select_Sites, courtesy of Esri.

\\EsriPress\GTKWebGIS\Data\Chapter8\Site_Selection.mxd, courtesy of Esri.

\\EsriPress\GTKWebGIS\Data\Chapter8\Assignment_Data\Scratch, courtesy of Esri.

\\EsriPress\GTKWebGIS\Chapter8\Data.gdb\Earthquakes, courtesy of USGS.

\\EsriPress\GTKWebGIS\Chapter8\Data.gdb\Hurricanes, courtesy of NOAA National Climatic Data
    Center.

\\EsriPress\GTKWebGIS\Data\Chapter8\Assignment_Data\natural_disasters.mxd, courtesy of Esri.

\\EsriPress\GTKWebGIS\Data\Chapter8\Assignment_Data\ExtractData.tbx, courtesy of Esri.

\\EsriPress\GTKWebGIS\Data\Chapter8\Assignment_Data\ExtractData.tbx\ExtractData, courtesy of
    Esri.

\\EsriPress\GTKWebGIS\Data\Chapter9\Lab_Data\m34117a2sw.tif, courtesy of USDA Farm Service
    Agency.

\\EsriPress\GTKWebGIS\Data\Chapter9\Assignment_Data\DTM_Petaluma.tfw, courtesy of NASA Grant
    NNX13AP69G, the University of Maryland, and the Sonoma Vegetation Mapping and LiDAR Program.

\\EsriPress\GTKWebGIS\Data\Chapter9\Assignment_Data\DTM_Petaluma.tif, courtesy of NASA Grant
    NNX13AP69G, the University of Maryland, and the Sonoma Vegetation Mapping and LiDAR Program.

\\EsriPress\GTKWebGIS\Data\Chapter9\Assignment_Data\DTM_Petaluma.tif.aux.xml, courtesy of
    NASA Grant NNX13AP69G, the University of Maryland, and the Sonoma Vegetation Mapping and
    LiDAR Program.

\\EsriPress\GTKWebGIS\Data\Chapter9\Assignment_Data\DTM_Petaluma.tif.ovr, courtesy of NASA
    Grant NNX13AP69G, the University of Maryland, and the Sonoma Vegetation Mapping and LiDAR
    Program.

\\EsriPress\GTKWebGIS\Data\Chapter10\2d3d_query.html, courtesy of Esri.

\\EsriPress\GTKWebGIS\Data\Chapter10\2d3d_query_key.html, courtesy of Esri.

\\EsriPress\GTKWebGIS\Data\Chapter10\2d3d_Views.html, courtesy of Esri.

\\EsriPress\GTKWebGIS\Data\Chapter10\2d3d_Views_key.html, courtesy of Esri.

\\EsriPress\GTKWebGIS\Data\Chapter10\2d3d_Views_legend.html, courtesy of Esri.

\\EsriPress\GTKWebGIS\Data\Chapter10\2d3d_Views_legend_key.html, courtesy of Esri.

\\EsriPress\GTKWebGIS\Data\Chapter10\load_basic_webmap.html, courtesy of Esri.

\\EsriPress\GTKWebGIS\Data\Chapter10\load_basic_webmap_key.html, courtesy of Esri.

\\EsriPress\GTKWebGIS\Data\Chapter10\load_basic_webscene.html, courtesy of Esri.

\\EsriPress\GTKWebGIS\Data\Chapter10\load_basic_webscene_key.html, courtesy of Esri.